**USDA**

United States
Department
of Agriculture

Forest Service

**Rocky Mountain
Research Station**

General Technical Report
RMRS-GTR-226

May 2009

# Sediment Transport Primer

# Estimating Bed-Material Transport

# in Gravel-bed Rivers

Peter Wilcock, John Pitlick, Yantao Cui

Wilcock, Peter; Pitlick, John; Cui, Yantao. 2009. **Sediment transport primer: estimating bed-material transport in gravel-bed rivers.** Gen. Tech. Rep. RMRS-GTR-226. Fort Collins, CO: U.S. Department of Agriculture, Forest Service, Rocky Mountain Research Station. 78 p.

## Abstract

This primer accompanies the release of BAGS, software developed to calculate sediment transport rate in gravel-bed rivers. BAGS and other programs facilitate calculation and can reduce some errors, but cannot ensure that calculations are accurate or relevant. This primer was written to help the software user define relevant and tractable problems, select appropriate input, and interpret and apply the results in a useful and reliable fashion. It presents general concepts, develops the fundamentals of transport modeling, and examines sources of error. It introduces the data needed and evaluates different options based on the available data. Advanced expertise is not required.

## The Authors

**Peter Wilcock,** Professor, Department of Geography and Environmental Engineer, Johns Hopkins University, Baltimore, MD.

**John Pitlick,** Professor, Department of Geography, University of Colorado, Boulder, CO.

**Yantao Cui,** Stillwater Sciences, Berkeley, CA.

You may order additional copies of this publication by sending your mailing information in label form through one of the following media. Please specify the publication title and series number.

**Fort Collins Service Center**

| | |
|---|---|
| **Telephone** | (970) 498-1392 |
| **FAX** | (970) 498-1122 |
| **E-mail** | rschneider@fs.fed.us |
| **Web site** | http://www.fs.fed.us/rm/publications |
| **Mailing address** | Publications Distribution |
| | Rocky Mountain Research Station |
| | 240 West Prospect Road |
| | Fort Collins, CO 80526 |

Rocky Mountain Research Station
240 W. Prospect Road
Fort Collins, Colorado 80526

# Disclaimer

## Download Information

The BAGS program, this primer, and a user's manual (Pitlick and others 2009) can be downloaded from: http://www.stream.fs.fed.us/publications/software.html. This publication may be updated as features and modeling capabilities are added to the program. Users may wish to periodically check the download site for the latest updates.

BAGS is supported by, and limited technical support is available from, the U.S. Forest Service, Watershed, Fish, Wildlife, Air, & Rare Plants Staff, Streams Systems Technology Center, Fort Collins, CO. The preferred method of contact for obtaining support is to send an e-mail to rmrs_stream@fs.fed.us requesting "BAGS Support" in the subject line.

U.S. Forest Service
Rocky Mountain Research Station
Stream Systems Technology Center
2150 Centre Ave., Bldg. A, Suite 368
Fort Collins, CO 80526-1891
(970) 295-5986

# Acknowledgments

The authors wish to thank the numerous Forest Service personnel and other users who tested earlier versions and provided useful suggestions for improving the program. We especially wish to thank Paul Bakke and John Buffington for critical review of the software and documentation. Efforts by the senior author in developing and testing many of the ideas in this primer were supported by the Science and Technology Program of the National Science Foundation via the National Center for Earth-surface Dynamics under the agreement Number EAR- 0120914. Finally, we wish to thank John Potyondy of the Stream Systems Technology Center for his leadership, support, and patience in making BAGS and its accompanying documentation a reality.

# Contents

# Chapter 1—Introduction

## Purpose and Goals

This primer accompanies BAGS (**B**edload **A**ssessment in **G**ravel-bedded **S**treams) software written to facilitate computation of sediment transport rates in gravel-bed rivers. BAGS provides a choice of different formulas and supports a range of different input information. It offers the option of using measured transport rates to calibrate a transport estimate. BAGS can calculate a transport rate for a single discharge or for a range of discharges. The "Manual for Computing Bed Load Transport Using BAGS (Bedload Assessment for Gravel-bed Streams) Software" (Pitlick and others 2009) provides a guide to the software, explaining the input, output, and operations step by step.

The purpose of this document is to provide background information to help you make intelligent use of sediment transport software and hopefully produce more accurate and useful estimates of transport rate. Although BAGS (or any other software) makes it easier to calculate transport rates, it cannot produce accurate estimates on its own. It can improve accuracy (mostly by reducing the chance of computational error), but it cannot prevent inaccuracy. In fact, by making the computations easier, BAGS and similar software makes it possible to produce inaccurate estimates (even wildly inaccurate estimates) very quickly and in great abundance.

Coming up with an accurate estimate of sediment transport rates in coarse-bedded rivers is not easy. If one simply plugs numbers into a transport formula, the error in the estimate can be enormous. To avoid this unpleasant situation, you need some understanding of how such errors can come about. This means you need to know something about transport models—what they are made of, how they are built, and how they work. The material presented in this manual, although somewhat detailed, is not particularly complicated. In fact, much of it is rather intuitive. Maybe you don't want to become an expert. But you should become an informed user—asking the right questions, making intelligent choices, developing reasonable interpretations, and evaluating useful alternatives when (as is usually the case) the amount of information you have is less than optimal. Although the manual contains some relatively detailed information, it does not presume that the reader has any particular experience estimating transport rates in rivers or in the supporting math and science. The primer is not intended for experts (although an expert may find useful material in it), but for practicing hydrologists, geomorphologists, ecologists, and engineers who have a need to estimate transport rates.

The remainder of Chapter 1 presents some general information, explaining sources of error in transport estimates, discussing the broader watershed context, and enumerating the various applications of sediment transport estimates. Chapter 2 provides a mini-course in sediment transport models for gravel-bed rivers, discussing the flow, nature of transport models, role of different measures of incipient grain motion, and importance of grain size. Chapter 3 draws from this information to lay out specifically the factors that give rise to error in transport estimates. Some background on the particular transport models used in BAGS is presented in Chapter 4 in order to help you evaluate which model may be appropriate for your application. Field data are needed for accurate transport estimates and we give some guidelines for data collection in Chapter 5. In Chapter 6, we evaluate the different options for making a transport estimate in terms of the available data. Because any transport estimate will have error, Chapter 7 presents a basis for estimating the magnitude of that error and suggests some strategies for handling that error in subsequent calculations and decisions.

Perhaps you are eager to begin making transport estimates. Before you skip ahead to the user's manual (or directly to the software itself), you should make sure that you are familiar with the general concepts described in the first section of Chapter 2 and the options available for estimating transport based on the data available, which are described in Chapter 6. If you work through the material in this primer, you can expect to understand why and how your transport estimate might be accurate or not, have some idea of the uncertainty in your estimate and what you might do to reduce it, and be able to consider alternative formulations that might better match the available information to the questions you are asking.

*Caveat emptor.* When calculating transport rates, it is very easy to be very wrong. Expertise in the transport business is only partly about understanding how to make reliable calculations. Another important part is recognizing situations in which the estimates are likely to be highly uncertain and figuring out how to reframe the question in a way that can be more reliably addressed. This primer will not make you an expert, but we hope that it can provide some context and answer key questions that will supplement your common sense and experience and help you pose and answer transport questions with some reliability. In some cases, an evaluation by someone with considerable experience and expertise would be advisable. In particular, these would include cases involving risk to highly valued instream and riparian resources and those with a potentially large supply of sediment. The latter could include stream design in regions with large sediment

USDA Forest Service RMRS-GTR-226. 2009.

supply and potential channel adjustments below large sediment inputs from dam removal, reservoir sluicing, forest fire, land-use change, or hillslope failures.

## Why it's Hard to Accurately Estimate Transport Rate

There are three primary challenges when using a formula to estimate transport rates. These will be discussed in detail in Chapter 3 after we have developed the basics of sediment transport modeling in Chapter 2. It will help to lay out the challenges at the beginning so you can keep the issues in mind as you go through the material. Here are the main culprits:

*The flow.* In many transport formulas, including those in BAGS, the flow is represented using the boundary shear stress $\tau$, the flow force acting per unit area of stream bed. Stress is not something we measure directly. Rather, we estimate it from the water discharge and geometry and hydraulic roughness of the stream channel. It is difficult to estimate the correct value of $\tau$ because it varies across and along the channel and only part of the flow force acting on the stream bed actually produces transport. So, we are trying to find only that part of $\tau$ that produces transport (we call it the grain stress) and a single value of grain stress that represents the variable distribution actually found in the channel. Figure 1.1 demonstrates the nature of this variability.

**Figure 1.1.** Henrieville Ck, Utah.

*The sediment.* Transport rate depends strongly on grain size. If we specify the wrong size in a transport formula, our estimated transport rate will be way off. Several factors make it difficult to specify the grain size. The range of sizes in a gravel bed is typically very broad. Fortunately, considerable progress has been made over the past couple of decades to develop models of mixed-size sediment transport. But, this wide range of sizes tends not to occur in a well-mixed bed with a simple planar configuration. Rather, the bed has topography and the sediment is sorted spatially by size and with depth into the bed (fig. 1.1). Even if we could thoroughly and accurately describe "the" grain size of a reach, we may not have the correct value to use in a transport formula because the sediment transported through the reach can be considerably different from that in the bed. Reliable use of a transport formula requires an interpretation of the nature of the stream reach. Is it in an adjusted steady state with the flow and transport (in which case the transport should be predictable as a function of bed grain size), or is it partly or fully nonalluvial (meaning that part or all of the sediment transport is derived from upstream reaches and does not reside within the reach)?

*The watershed.* Because questions of sediment supply and alluvial adjustment intrude on the calculation of transport rates, an understanding of the dynamics and history of your watershed is needed in order to choose an appropriate study reach for analysis and to provide a basis for evaluating the results. Watershed factors are closely related to the sediment problem because they influence the sediment supply. Is it changing in time or along the channel? Is it substantially different from what is found in the stream bed? An example would be a stream reach downstream of a jam of large woody debris. Even a single tree fall can trap a large fraction of the sediment supply. This will change the transport and bed composition in the reach in which you are working.

The underlying reason why uncertainty in transport estimates is so large is that the formulas (actually, the underlying physical mechanisms) are strongly *nonlinear*. The significance of this is that if you are off a little bit on the input, the calculated transport rates can be way off. If your input is off by 50 percent, your calculated transport rate will be off by more (sometimes much more) than 50 percent. It is very easy to predict large transport rates when little transport actually occurs, or to predict no transport when the actual transport is quite large.

If the challenges involved in developing a reliable transport estimate seem a bit daunting, they should. They are. Even with data from a field visit where you conduct a cross-section survey, collect a pebble count, and estimate the channel slope, you cannot assume you will have a transport estimate of useable accuracy. BAGS will make it easier to estimate transport rates, but it won't make the estimates more accurate. That is up to you. There are a variety of things you can do

4

to improve the accuracy of your transport estimate and effectively accommodate uncertainty in addressing the broader questions that motivated you to estimate the transport rate in the first place. This is why we wrote this primer.

We also provide some guidance on choosing the location and data for making reliable transport estimates. But your job is not finished when you type some input and get a transport estimate from BAGS. You have to critically evaluate the outcome, taking into account channel and watershed dynamics and making use of common sense observations. With a sound understanding of transport basics, you can assess the uncertainty in your estimated transport rate and decide whether it is acceptable or you need to take steps to improve the estimate or redefine the problem in a way that accommodates the uncertainty. The goal of this primer is to explain the tools needed for these tasks and make you a critical and effective user of the sediment transport software.

## Watershed Context of Sediment Transport Problems

Every stream has a history. This history is likely to have a dominant and persistent influence on the sediment transport rates. Every stream has a watershed, with hydrologic, geologic, and biologic components. The nature of the watershed, timing and location of any disturbances within the watershed, and time needed for these disturbances to work their way through the watershed will all have a dominant influence on water and sediment supply, stream characteristics, and transport rates at the particular location where you would like to develop a transport estimate.

We can't cover watershed hydrology and geomorphology or fluvial geomorphology in this primer, but we cannot ignore this essential topic. In most cases, it is hard to imagine that a transport estimate made in the absence of a sound understanding of watershed history and dynamics would be of much use at all. Often, the most accurate (if imprecise) estimate of transport rate—and certainly any estimate of the trends in transport rates—will be derived from a description of slope, dimension, runoff, and land use throughout the watershed. Together, these provide an indication of whether the transport in your reach may be increasing or decreasing, coarsening or fining. A sound understanding of watershed history and context is needed to develop and evaluate plausible estimates of sediment transport rate (Reid and Dunne 1996, 2003). Because a sediment transport estimate is usually just one component of a broader study, an understanding of the watershed is likely to be key in addressing the larger issues you are grappling with.

Although there may often be limited data available for a particular stream reach, useful information for assembling the story of your watershed can often be collected quite easily. Extensive flow records for comparable streams can often be retrieved from the Internet (http://waterdata.usgs.gov/nwis) and aerial photograph coverage extending back 70 to 80 years is now commonly available (http://edc.usgs.gov/,http://www.archives.gov/publications/general-info-leaflets/26.html#aerial2). County soil surveys can provide extensive and detailed information on the soils, geomorphology, and drainage of the watershed (http://soils.usda.gov/survey/). State and county planning offices often have land-use records available on line. Previous watershed studies may be available from the U.S. Forest Service, TMDL studies, and the EPA Watershed Assessment Database (http://www.epa.gov/waters). This information, combined with a broad understanding of historical channel adjustments can provide a sound context, with modest effort, for your transport estimate (for example, Gilvear and Bryant 2003; Jacobson and Coleman 1986; Trimble 1998).

Historical records will not provide precise quantitative information on the historical supply of water and sediment to your reach, but an accurate assessment of the *relative* trends in water and sediment supply may be possible and sufficient to provide a useful assessment of past and future channel changes. A basis for making such assessments was suggested by Lane (1955), who proposed a simple balance between slope and the supply of water and sediment:

$$Q_s D \sim QS \qquad (1.1)$$

where $Q_s$ is sediment supply, $D$ is the grain size of the sediment, $Q$ is water discharge, and $S$ is channel slope. This relation was illustrated by Borland (1960) in a form that memorably captures the interaction between water and sediment supply and channel aggradation/degradation (fig. 1.2). Although evocative, neither the figure nor Eq. 1.1 supports quantitative analysis because the nature of the function is not specified. As a result, it is also indeterminate in some important cases, such as when the sediment load increases and becomes finer-grained.

The stable channel balance can be quantified if appropriate relations for flow and transport are specified. A simple analysis by Henderson (1966) is useful, but has received surprisingly little attention. Henderson combined the Einstein-Brown transport formula with the Chezy flow resistance formula, and momentum and mass conservation for steady uniform flow, into a single proportionality:

$$q_s D^{3/2} \propto (qS)^2 \qquad (1.2)$$

where $q_s$ and $q$ are sediment transport rate and water discharge per unit width. For the purpose of interpreting past or future channel change, Eq. 1.2 is more usefully solved for $S$:

$$S \propto \frac{\sqrt{q_s}\,D^{3/4}}{q} \qquad (1.3)$$

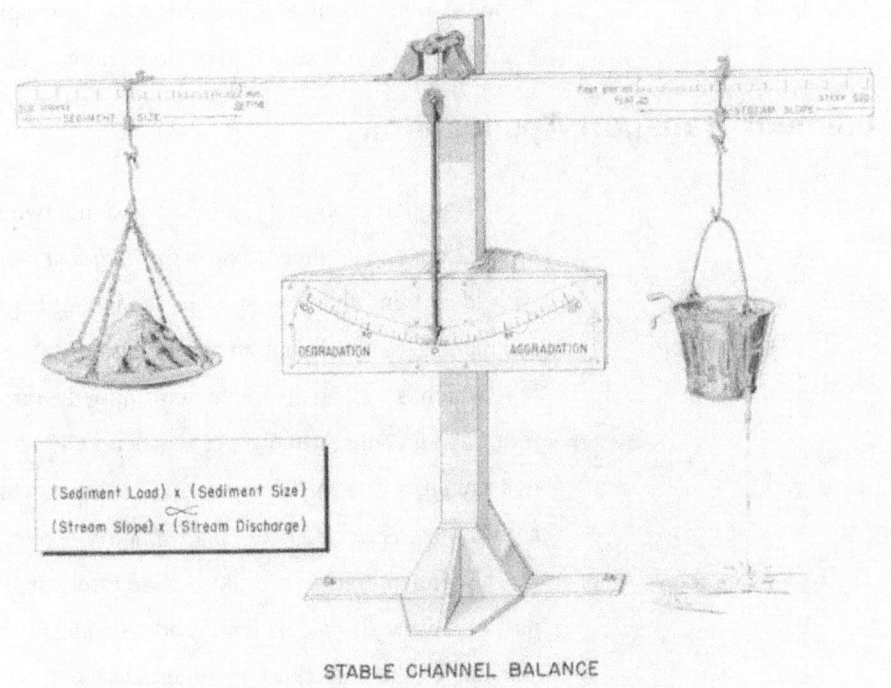

**Figure 1.2.** The Lane/Borland stable channel stability relation (Borland 1960).

STABLE CHANNEL BALANCE

Writing Eq. 1.3 twice, for the same reach at two different time periods, and taking the ratio:

$$\frac{S_2}{S_1} = \left(\frac{q_{s2}}{q_{s1}}\right)^{1/2} \left(\frac{q_1}{q_2}\right)\left(\frac{D_2}{D_1}\right)^{3/4} \qquad (1.4)$$

Eq. 1.4 can be applied to the evaluation of channel change if $D$ and $q_s$ are the grain size and rate of sediment supply to the reach and $q$ to be the water supply to the reach. In this case, $S$ in Eqs. 1.3 and 1.4 can be interpreted as the slope necessary to transport the sediment supplied (at rate $q_s$) with the available flow $q$. An increase in $S$ ($S_2/S_1 > 1$) is not likely to be associated with a large increase in bed slope (which would generally take a very long time), but rather indicates bed aggradation (as in fig. 1.2), or, more accurately, a tendency for the channel to accumulate sediment under the new regime. A decrease in $S$ represents degradation, or a tendency for the channel to evacuate sediment under the new regime, thus linking back to Lane's balance. In cases where little reliable information on water and sediment supply is available (for example, perhaps only the sign and approximate magnitude of changes in $q$ and $qs$ are well known), Eq. 1.4 can nonetheless provide a useful estimate of the tendency of the channel to store or evacuate sediment. Such an estimate may be at least as reliable (and perhaps more reliable) as that provided by more detailed calculations based on highly uncertain boundary conditions. Certainly, any predictions based on detailed calculations should be consistent with an estimate based on Eq. 1.4 and the accumulated knowledge about channel change in the region. Clark and Wilcock (2000) used this relation to evaluate channel adjustments in response to historical land use and sediment supply trends in Puerto Rico. Schmidt and Wilcock (2008) used it to evaluate downstream impacts of dams.

# Sediment Transport Applications

Transport problems can be divided into two broad classes, each with different applications and methods. One is the *incipient motion* problem, which is concerned with identifying the flow at which sediment begins moving or identifying which sediment sizes are in motion at a given flow. The other is the *transport rate* problem, which is concerned with determining the rate at which sediment is transported past a certain point, usually a cross-section. If a flow is sufficient to move sediment in a stream, it is termed *competent*. The rate at which the stream moves sediment at a given flow is termed *transport capacity*.

Sediment transport estimates are rarely an end in themselves, but instead are part of a suite of calculations used to address a larger problem. A sound understanding of the objectives and alternatives of the broader problem can help guide decisions about approaches and the effort appropriate for a transport analysis. This is particularly important because sediment transport estimates generally have considerable uncertainty and, by placing the transport estimate within its broader context, it may be possible to find ways to reframe the question to best match the available data. For example, if you are interested in the future condition of a stream reach, the *difference* between the transport capacity today and in the future, and the *difference* between that transport capacity and the rate of sediment supply to the reach are of more importance than the actual rate of transport. This is because the difference determines the amount of sediment that will be stored or evacuated from the reach, producing channel change. Often, a difference can be calculated with more accuracy than the individual values themselves. This will be discussed further in Chapter 6.

## Incipient Motion Problems

One incipient motion problem is to determine the flow at which any grains on the bed and banks of a stream will be transported. If a channel is intended to remain static at a design flow, the designer is interested in finding the dimensions and grain size of a channel that are as efficient as possible (minimizing the amount of excavation) without entraining any grains from the bed or banks (for example, Henderson 1966). These ideas are also applied in urban stream design and to channels below dams because, in both cases, there may be little or no sediment supply available to replace any grains that are entrained. Thus, any transport will lead to channel enlargement and a static or threshold channel is sought.

A related incipient motion problem is determining the frequency with which bed or bank sediment is mobilized, given the flood frequency and channel properties. This can be useful for defining the ecologic regime of a channel, particularly the frequency and timing of benthic disturbance (Haschenburger and Wilcock 2003).

A more detailed incipient motion problem concerns the proportion of the stream bed that is entrained at a particular discharge. Some floods may produce transport for only a portion of the grains on the bed, a condition termed *partial transport* (Wilcock and McArdell 1997). The proportion of the bed entrained is relevant for defining the extent of benthic disturbance and the effectiveness of flows in accessing the bed substrate needed for flushing fine sediment from spawning and rearing gravels.

## Estimating Sediment Loads

Estimates of sediment transport rate are needed to determine the annual sediment load, calculate sediment budgets, and estimate quantities of gravel extraction or augmentation. These estimates are also needed to assess stream response to changes in water and sediment supply (for example, from fires, landslides, forest harvest, urbanization, or reservoir flushing) and determine the impact of these changes on receiving waters (for example, reservoir filling and downstream water quality impacts).

We also need to know rates of sediment transport in order to predict channel change. As Eq. 1.1 indicates, stream channel change depends on both water and sediment supply. Changes in sediment transport rate along a channel are balanced by bed aggradation/degradation and bank erosion. Anticipating these changes and designing channels that will successfully convey the supplied sediment load with the available water is the goal of stable channel design.

## Identifying the Correct Sediment Transport Problem

It is common for the wrong sediment transport principle—incipient motion versus transport rate—to be applied to a problem. For example, calculation of transport rates is inappropriate if the problem concerns determining the dimensions of a threshold channel (a channel in which none of the bed and bank sediment should move). It is also inappropriate if the question concerns simply the frequency of bed disturbance. Although a transport calculation includes an estimate of incipient motion (because this defines the intercept in a transport relation) and thus can indicate whether sediment moves or not at a given flow, what is of greater concern in a threshold channel analysis is the degree to which the flow falls below the threshold of motion. This difference indicates the extent to which a channel design can be changed, perhaps at considerable savings, while still meeting design requirements.

For existing channels, there are simple and inexpensive field methods for determining the discharge producing incipient motion (for example, placing painted rocks on the stream bed and observing if they were displaced by different discharges).

More serious problems can ensue if a transport rate problem is mistaken for an incipient motion problem. Commonly, a stream is assumed to be capable of transporting its sediment supply if its bankfull discharge can be shown to be *competent* (that is, the bankfull discharge is calculated to exceed the critical discharge for incipient motion of grains on the bed). Channel change is determined by the balance of sediment supply and the transport *capacity* of the reach. A reach may be competent at bankfull flow, but its transport capacity may be smaller than the rate at which sediment is supplied. In this case, sediment will deposit in the reach, which may be expected to lead to the growth and migration of gravel bars and associated erosion of channel banks. Conversely, a reach may be competent at bankfull flow, but its transport capacity may be larger than the rate at which sediment is supplied. In this case, sediment will be evacuated from the reach, which may be expected to lead to bed incision and armoring.

## Two Constraints

Two overarching constraints bound any approach to estimating transport rates in gravel-bed rivers. These are the spatial and temporal variability of the transport process itself and the sparse information that is typically available for developing an estimate of bed-material transport. The transport of bed material in gravel-bed rivers is driven by strongly nonlinear relations controlled by local values of flow velocity and bed material grain size. For the purpose of developing a transport estimate from field observations, the large variability requires a dense array of long duration samples for adequate accuracy. For the purpose of developing estimates from a transport formula, the large variability, combined with the steep nonlinear relations governing transport, make predictions based on spatial and temporal averages inaccurate. The second constraint—sparse information—is directly related to the first. If there were little variability in the transport, only a few observations would provide a representative sample. Sparse information strongly affects our ability to estimate transport from a formula. Models that are sensitive to local details of flow and bed material (for example, mixed-size transport models using many size fractions) require abundant local information for accurate predictions. This information is seldom available for an existing channel and can be specified for a design reach only at the time of construction. Transport and sediment supply in subsequent transport events will alter the composition and topography of the stream bed.

# Chapter 2—Introduction to Transport Modeling

## General Concepts

### Grain Size

In sediment transport, size matters in two ways. First, larger grains are harder to transport than smaller grains. It takes less flow to move a sand grain than a boulder. We can call this an *absolute* size effect. Second, smaller grains within a mixture of sizes tend to be harder to move than they would be in a uni-size bed, and larger grains tend to be easier to move when in a mixture of sizes. We can call this a *relative* size effect. Relative size matters in gravel-bed rivers because the bed usually contains a wide range of sizes.

We need some nomenclature for describing grain size. Because of the wide range of sizes, we use a *geometric* scale rather than an *arithmetic* scale. (You might think of a 102-mm grain as about the same size as a 101-mm grain, and a 2-mm grain as much bigger than 1-mm grain. If so, you are thinking geometrically. On an arithmetic scale, the *difference* in size is the same in both cases [1-mm]. On a geometric scale, the 2-mm grain is twice as big as the 1-mm grain.) The geometric scale we use for grain size is based on powers of two. Although originally defined as the $\Phi$ (phi) scale, where grain size $D$ in mm is $D = 2^{-\Phi}$, in gravel-bed rivers the $\psi$ (psi) scale is used, where $\psi = -\Phi$, or $D = 2^{\psi}$. Table 2.1 presents common names for different grain size classes.

**Table 2.1.** Common grain size classes.

| (mm) | | | Size class |
|---|---|---|---|
| – | to | <0.002 | clay |
| 0.002 | to | 0.004 | vf silt |
| 0.004 | to | 0.008 | f silt |
| 0.008 | to | 0.016 | m silt |
| 0.016 | to | 0.031 | c silt |
| 0.031 | to | 0.063 | vc silt |
| 0.063 | to | 0.125 | vf sand |
| 0.125 | to | 0.25 | f sand |
| 0.25 | to | 0.5 | m sand |
| 0.5 | to | 1 | c sand |
| 1 | to | 2 | vc sand |
| 2 | to | 4 | vf gravel |
| 4 | to | 8 | f gravel |
| 8 | to | 16 | m gravel |
| 16 | to | 32 | c gravel |
| 32 | to | 64 | vc gravel |
| 64 | to | 128 | f cobble |
| 128 | to | 256 | c cobble |
| | >256 | | boulder |

(vf: very fine; f: fine; m: medium; c: coarse; vc: very coarse).

Even a cursory examination of real streams demonstrates that the range of sizes in the bed is typically very large. Although a standard nomenclature for mixtures of sizes in gravel beds is not well developed (as it is for soils, for example), a simple means of describing a size mixture is to use the name (for example, gravel or cobble) representing the size class containing the largest proportion of the mixture and to modify this name using another size class containing a substantial amount of sediment (for example, a sandy gravel or a cobbly gravel). Buffington and Montgomery (1999a) provide more information on classifying fluvial sediment.

Grain-size distributions are commonly plotted as cumulative curves, giving percent finer versus grain size. The sediment shown in figure 2.1 has 10 percent finer than 4 mm, 30 percent finer than 8 mm, 50 percent finer than 16 mm, 70 percent finer than 32 mm, and 90 percent finer than 64 mm, all by weight (or volume). We use "percent finer" to describe characteristic grain sizes, usually presented as $D_{xx}$ with xx being an integer between 1 and 99, such that xx percent of the sediment (by weight or volume) is finer than $D_{xx}$. For example, $D_{90}$ represents that 90 percent of the sediment is finer than $D_{90}$ and $D_{50}$ is the median grain size. $D_{50}$ and $D_{90}$ values are 16 mm and 64 mm, respectively, in the grain size distribution shown in figure 2.1. The hydraulic roughness of a stream bed is often represented using a coarser grain size (for example, $D_{90}$ or $D_{84}$) and the transport rate is often calculated relative to its median size $D_{50}$.

To calculate the transport rate of different sizes within a mixture, we use the proportion in different size fractions. Let $D_1$, $D_2$, ..., $D_{N+1}$ be the grain sizes with associated percent finer values of $P_{f1}$, $P_{f2}$, ..., $P_{fN+1}$. Thus, N size ranges between

**Figure 2.1.** Example of a cumulative grain-size distribution curve.

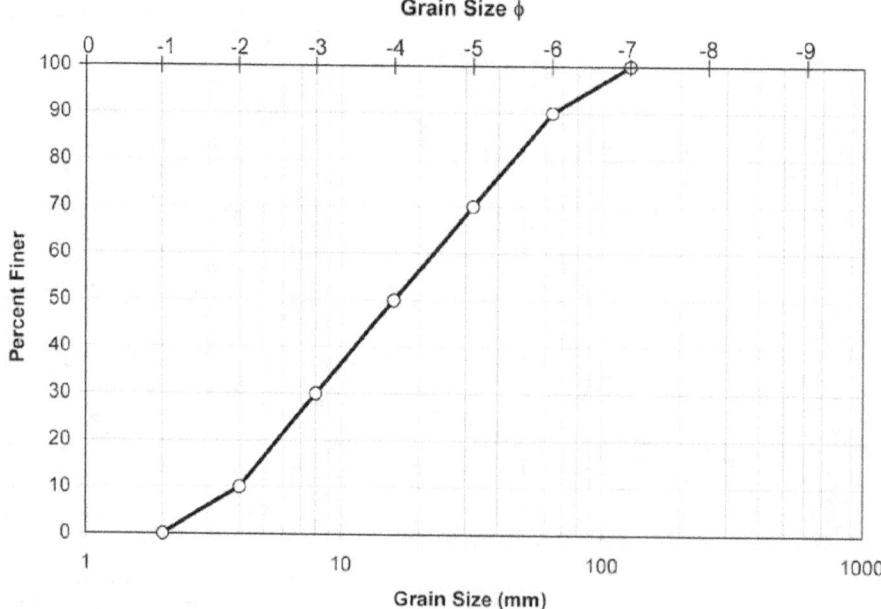

USDA Forest Service RMRS-GTR-226. 2009.

$D_1$ and $D_2$, $D_2$ and $D_3$, ..., $D_N$ and $D_{N+1}$, will have associated volumetric fractions $F_1$, $F_2$, ..., and $F_N$. The mean size of each group and the associated volumetric fraction are calculated as:

$$\overline{D_i} = \sqrt{D_i D_{i+1}} \ , \quad \overline{\Psi_i} = \frac{\Psi_i + \Psi_{i+1}}{2} \ , \quad F_i = \frac{\left| P_{fi+1} - P_{fi} \right|}{100} \qquad \text{(2.1 a,b,c)}$$

In addition to the median grain size, we represent the center of a size distribution using the mean:

$$\overline{\Psi} = \sum_{i=1}^{N} \overline{\Psi_i} F_i \ , \quad D_g = 2^{\overline{\Psi}} \qquad \text{(2.2 a,b)}$$

where $\overline{\psi}$ is the arithmetic mean in the $y$ scale and $D_g$ is the geometric mean. The spread of the size distribution is represented by the standard deviation:

$$\sigma_{\Psi} = \sqrt{\sum_{i=1}^{N} \left( \overline{\Psi_i} - \overline{\Psi} \right)^2 F_i} \ , \quad \sigma_g = 2^{\sigma_{\Psi}} \qquad \text{(2.3 c,d)}$$

where $s_y$ is the arithmetic standard deviation in the $y$ scale and $s_g$ is the geometric standard deviation in mm. For the example, in figure 2.1, $\overline{\Psi} = 4$, $D_g = 16$ mm, $s = 2.25$, and $s_g = 4.76$. Although this example has identical $D_g$ and $D_{50}$ values, they are generally different from each other. Note that the range of sizes within one standard deviation of the mean is found arithmetically on the $y$ scale as $\overline{\Psi}$ $\pm s$ (from $y = 1.75$ to $y = 6.25$) and geometrically on the $D$ scale (from $D_g/s = 3.36$ mm to $D_g s = 76.1$ mm).

One more descriptor of gravel beds is useful. We can think of a gravel bed as being formed by a three-dimensional framework of grains. The pore spaces between these grains may be empty, or they may contain finer sediments, particularly sand. As long as the proportion of sand is smaller than about 25 percent, nearly all of the bed is composed of gravel grains in contact with each other. We call this a *framework-supported* bed. If the proportion of sand increases further, some of the gravel grains are no longer fully supported by contacts with other gravel grains. With enough sand (more than roughly 40 percent), few gravel grains remain in contact. Rather, they are supported by a matrix of finer sediment and we refer to this as a *matrix-supported* bed. As we will discuss later, gravel in a matrix-supported bed tends to be transported at much higher rates.

## Surface or Subsurface?

In addition to sorting by grain size across and along the streambed surface, gravel beds tend to also exhibit vertical sorting, wherein the surface of the streambed is coarser than the underlying material. This is referred to as bed armoring (Parker and Sutherland 1990). In the transport literature, the material below the bed surface is referred to as both subsurface and substrate (as distinct from using the term substrate to refer to the channel bottom more generally). Vertical size sorting introduces a problem: should we use surface or subsurface grain size in a transport formula?

A variety of studies have shown that the transported load, integrated over a range of flows, will be finer than the surface and closer in size to the bed substrate (Church and Hassan 2002; Lisle 1995). Many transport formulas are based on flume experiments and have been developed using the grain size of the bulk sediment mix. Because the bulk mix approximates the substrate, not the surface, a substrate grain size is most appropriate when using these formulas. Unfortunately, this approach poses a serious problem. The transport at any moment must depend on the sizes available for transport on the bed surface. But the composition of the bed surface will depend on the history of flow and the sediment supply. Different streams have different histories and two streams with the same substrate grain size are not likely to have the same surface grain size. But a substrate-based transport formula would predict the same transport rates in each case.

If the transport is predicted in terms of the bed substrate grain size, the connection between the bed and transport is made through the bed surface, whose composition depends not only on the immediate physical processes of transport, but also on the sediment supply and the preexisting bed structure and composition. It seems unreasonable to expect a transport formula to account for bed sorting in response to variable initial and boundary conditions. The appropriate approach is to define the transport relative to the composition of the bed surface. It is the absence of coupled surface and transport observations that requires transport models to be referenced to the substrate or bulk size distribution of the bed. Recent laboratory experiments have now provided such data (Wilcock and others 2001) and surface-based transport formulas can now be tested against data.

Transport formulas for mixed-size sediments predict larger transport rates for finer fractions—the predictions are size-selective. Thus, the observation that transport through a reach is finer than the bed surface does not necessarily indicate that the reach is out of equilibrium.

## What Transport Looks Like

The sediment in gravel beds is immobile most of the time. Flows sufficient to move sediment generally occur during only a small fraction of the year and many of these transport only sand over a bed of immobile gravel. Active transport of the framework grains occurs in larger flows, which might occur a few times per year or less. Even when these grains are actively transported, most of the grains on the bed surface are not moving most of the time. Grains are observed to rock back and forth and occasionally individual coarse grains will roll, slide, or hop along the bed. Bed load transport in gravel-bed streams is an intermittent, spatially variable, and stochastic process. This is nicely illustrated in video of transport in gravel-bed streams (for example, "Viewing Bedload Movement in a Mountain Gravel-bed Stream" at http://www.stream. fs.fed.us/publications/videos.html; see also video available at http://www.public.asu.edu/~mschmeec/).

Additionally, after floods that move considerable amounts of sediment, there may be parts of a gravel bed that remain at least partly undisturbed. For example, one can measure large transport rates that include all sizes found in the bed, but still find that some grains on the bed surface never moved. Recall that we defined this as *partial transport*—the condition in which some grain move and others do not (Wilcock and McArdell 1993, 1997). The occurrence of partial transport can sometimes be easily observed in the field if the exposed parts of bed-surface grains develop a chemical or biological stain during low flow periods. After a transporting event, partial transport will be evident in regions of the bed showing few fresh surfaces. The flow at which all the grains of a particular size are moved is larger for larger grains, and the magnitude of a flood producing complete mobilization of the bed surface may be very large, exceeding a five- or 10-year recurrence interval (Church and Hassan 2002; Haschenburger and Wilcock 2003). The proportion of a size fraction that remains inactive over a flood will have an influence on transport rates and is immediately important for estimating exposure of the bed substrate to the flushing action of high flows.

## Transport Mechanisms and Sources

Sediment transport is often separated into two classes based on the mechanism by which grains move: (1) *bed load*, wherein grains move along or near the bed by sliding, rolling, or hopping and (2) *suspended load*, wherein grains are picked up off the bed and move through the water column in generally wavy paths defined by turbulent eddies in the flow. In many streams, grains smaller than about 1/8 mm tend to always travel in suspension, grains coarser than about 8 mm tend to always travel as bed load, and grains in between these sizes travel as either bed load or suspended load, depending on the strength of the flow (fig. 2.2). We divide transport into these categories because the distinction helps to develop an understanding of how transport works and what controls it.

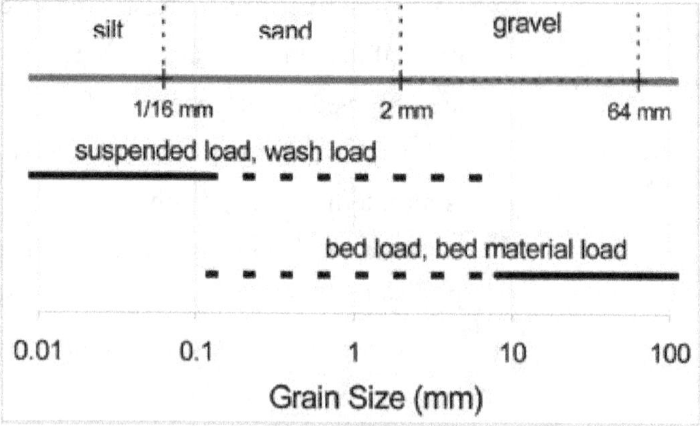

**Figure 2.2.** Grain sizes associated with bed load, bed-material load, suspended load, and wash load.

Sediment transport can be organized in another way based on the source of the grains: (1) *bed material load*, which is composed of grains found in the stream bed; and (2) *wash load*, which is composed of finer grains found in only small (less than a percent or two) amounts in the bed. The sources of wash load grains are either the channel banks or the drainage area contributing runoff to the stream. Wash load grains tend to be very small (clays and silts and sometimes fine sands) and, hence, have a small settling velocity. Once introduced into the channel, wash-load grains are kept in suspension by the flow turbulence and essentially pass straight through the stream with negligible deposition or interaction with the bed.

The boundary between bed load and suspended load is not sharp and depends on the flow strength. Consider a stream with a mixed bed material of sand and gravel. At moderate flows, the sand in the bed may travel as bed load. As flow increases, the sand may begin moving partly or entirely in suspension. Even when traveling in suspension, much of this sediment (particularly the coarse sand) may travel very close to the bed, down among the coarser gravel grains in the bed. That makes it very difficult to sample the suspended load in these streams or, for that matter, to even distinguish between bed load and suspended load. This difficulty is one reason why we focus in this manual on *bed material load* rather than bed load and suspended load. Another reason is one of simplicity: the bed material in a stream can be defined and measured. We are interested in its transport rate and should invoke the alternative classification—based on transport mechanisms—only if it helps us reach our goal of estimating transport rates.

When we use a transport formula, we attempt to predict the transport rate in terms of the channel hydraulics and the bed grain size. We don't try that with wash load because its transport rate depends on the rate at which these fine sediments are supplied to the stream rather than properties of the flow and stream bed. Now, it turns out that bed material can behave at least partially like wash load in the sense that the sediment passing through a reach may be entrained from the bed somewhere upstream. The reach may function more like a pipe that simply passes the upstream sediment supply versus a stream bed that actively exchanges sediment between the bed and the transport. If we apply a transport formula to a pipe-like reach, we will calculate negligible transport, even though there might be a lot of sediment passing through it. Detecting such situations is essential for accurate transport estimates from formulas. Using measured transport rates to calibrate a transport formula goes a long way toward addressing this problem. We discuss this problem in the next section and return to it in Chapter 3—The Sediment Problem.

An important concept regarding bed material load is the effect of sediment supply on transport rates. If the supply of wash load range is increased, we will observe an increase in the wash load, but the transport rates of the coarser grain sizes—comprising the bed material—will remain unchanged (unless we add so much wash load material that the flow turns into a thick slurry resembling pea soup). In contrast, if the supply of bed material is changed, we expect that the bed composition will change as well and, therefore, the transport rates of the bed material will also change. For example, if the supply of coarse sand to a gravel-bed stream were increased (as from land clearing or a forest fire), then we would expect the amount of sand in the bed to increase. By increasing the sand content and thereby reducing the gravel content of the bed, we might expect that sand transport rates would increase and gravel transport rates would decrease. It turns out that increasing the sand content increases the transport rate of *both* sand and gravel (Wilcock and others 2001). The important distinction here is that altering the supply of sediment in one size range of the *bed material* will alter the bed composition and the transport rates, whereas altering the supply of sediment in the size range of *wash load* will have negligible effect on the bed composition and bed material load. This distinction may seem picky at this point, but it is important in understanding transport rates and channel change in response to changes in sediment supply to a stream channel.

It is useful to distinguish between different sizes of bed material. Fine bed material load typically consists of medium to coarse sand and, in many cases, pea gravel, which can move as either bed load or suspended load. When in suspension, the grain trajectory is typically within a near-bed region where the flow is locally disturbed by wakes shed from the larger grains in the bed. Fine bed material exists in the interstices of the bed and in stripes and low dunes at larger concentrations. The near-bed suspension of the fine bed material cannot be sampled with conventional suspended sediment samplers and models for predicting its rate of transport are incomplete. Coarse bed material forms the framework of the river bed. Its motion is almost exclusively as bed load. Displacements of individual grains are typically rare and difficult to sample with conventional methods. In some streams, we can distinguish another, yet coarser fraction, typically in the boulder size class, which is immobile at typical high flows. Although not contributing to the transport, these grains do contribute to the hydraulic roughness of the channel. Their effect must be included in any flow calculation.

Bed material transport is the basic engine of fluvial geomorphology. The balance between its supply and rate of transport in a stream channel governs bed scour and aggradation, channel topography and flow patterns, and the subsequent erosion and construction of bars, bends, banks, and floodplains.

## Sediment Supply Versus Transport Capacity

The transport rate in a channel—the quantity calculated by BAGS—is termed the *transport capacity*. Any imbalance between the transport capacity and the sediment supply rate determines the amount of sediment deposited or eroded in the channel and the associated channel change. It can take time to produce channel change, particularly if the rates of transport are small. Different types of channel adjustment require the transport of different amounts of sediment and thus can be anticipated as occurring in a given order. Changes may be expected first in the grain size of the stream bed, followed by construction or removal of in-channel bars, streambed incision or aggradation, and bank erosion. Changes in stream planform and, finally, channel slope require the rearrangement of large quantities of sediment and take much longer (Parker 1990a).

The distinction between sediment supply and transport capacity highlights two important problems with estimating transport rates. The first is more relevant to estimating transport rates from field measurements and the second to calculating transport rates from a formula. First, minor changes in sediment storage (slight aggradation or degradation) may strongly influence transport rates in a reach. For example, a fallen tree may trap all of the sediment transport in a stream with relatively small transport rates. Somebody unfortunate enough to measure transport rates downstream of the tree fall would observe little or no transport, producing a very misleading record. Although this case is rather obvious, small amounts of bed aggradation or degradation upstream or within a sampling reach could result in the trapping or release of a large fraction of the sediment supply. It is always a useful exercise to compare measured or predicted transport rates against the amount of aggradation or deposition those rates could produce. For example, if one calculated an annual sediment load for a reach, it could be useful to determine the change in bed thickness that would result if a large fraction of this sediment were evenly deposited over the reach. If the change in elevation is small, it is inadvisable to presume much precision in the estimated transport rates.

A second problem concerns the grain size to be used in a transport formula. If a reach is fully alluvial and at equilibrium, such that the channel is formed of the material the stream is transporting and the transport rates in and out of the reach are balanced over periods of a storm or longer, one could reasonably measure the grain size in a reach and insert this into a transport formula. If, however, the reach is not fully alluvial or in equilibrium, the sediment in transport may be substantially different in size from that in the channel bed. An extreme example would be a coarse, armored stream below a dam, in a reach just below a tributary supplying finer grain sediment. If there is sufficient flow to transport the finer sediment in the mainstem, the grain size of the transport may be entirely different from that of the coarse armored bed. Thus, it would not be possible to predict the transport rate using the grain size of the bed. Although this is an extreme case, it does illustrate that one cannot presume to predict the transport rate using the grain size of the bed. It must be established that the bed material has adjusted to be in a steady state with the sediment supply.

The nature of the sediment supply problem will vary with location in a watershed. In headwater reaches, stream channels are generally more closely coupled with the adjacent hillslopes. A larger fraction of the bed material may have been introduced via local hillslope processes than would be the case lower in the watershed. If some of this material is very coarse and effectively immobile, the transport capacity estimated from a measurement of bed material grain size may be in error.

## Sediment Rating Curves

Most practical sediment transport problems require definition of the sediment transport rate $Q_s$ as a function of water discharge, $Q$. A relation giving $Q_s$ as a function of $Q$ is called a **sediment rating curve**. A sediment rating curve is often represented as a power function:

$$Q_s = aQ^b \tag{2.4}$$

where, in the United States, $Q_s$ is in units of tons per day and $Q$ is in units of ft$^3$/s, or cfs. Preferable units would be kg/hr or Mg/day and m$^3$/s.

An essential part of developing a transport model is developing a basis for scaling or representing the discharge $Q$. Because most applications require a prediction of transport as a function of discharge, the obvious step is to try to develop a model based directly on $Q$. This model is not likely to be general. It is quite unlikely that, say, 100 cfs would produce the same transport rate in a small creek compared to a very large river (a km wide or more). Thus, the coefficient $a$ in Eq. 2.4 may be expected to vary quite widely among different rivers. Further, differences in channel size, shape, slope, roughness, and bed material will cause the rate at which $Q_s$ varies with $Q$ to differ widely, indicating that the exponent $b$ in Eq. 2.4 would also take a wide range of values for different rivers.

A dimensionless sediment rating curve has been proposed in which $Q_s$ and $Q$ are divided by their values measured at flows close to bankfull (Rosgen 2007). Assuming that the coefficient $a$ does not vary with $Q$, this has the desirable effect of eliminating it from the relation, leaving only the exponent $b$ to be specified. Unfortunately, the exponent $b$ varies widely from one river to another so the model is not predictive. Use of a single value of $b$ (a value of 2.2 is suggested by Rosgen 2007) will lead to large errors in predicted transport rate and cannot be recommended. Barry and others (2004, 2005) explore the variation of $a$ and $b$ using a large field data set.

# The Flow

A measure of flow strength that has been found to provide a generalized description of transport rate is the bed shear stress, $\tau$. Stress is a force per area: in this case, the shear force exerted by the flowing water on an area of the bed. Reasonably, the transport should depend on the fluid force applied to the bed, but estimating $\tau$ is difficult.

## Non-Uniform and Unsteady Flow

Flow that does not vary in time is described as steady. Flow that does not vary alongstream is termed uniform. For steady, uniform flow, the stress acting on the bed is:

$$\tau_0 = \rho g R S \tag{2.5}$$

where $R$ is the hydraulic radius, given by ratio of flow area $A$ to wetted perimeter $P$, and $S$ is the bed slope. We use rise over run, or tana, where a is the bed slope angle used to calculate bed slope. (Strictly, the correct value of slope to use in Eq. 2.5 is sina, but for the slopes typical of rivers, sina nearly equals tana.) Although Eq. 2.5 uses $R$, it is often referred to as the *depth-slope product*. In channels with a ratio of width to depth ($B/h$) greater than about 20, $R \approx h$ within 10 percent.

No natural flow is perfectly uniform or steady. For the more complex but realistic case in which the flow can accelerate in both time (discharge changes) and in space (flow is non-uniform), the boundary stress is given by the one-dimensional St. Venant equation:

$$\tau_0 = \rho g R \left( S - \frac{\partial h}{\partial x} - \frac{U}{g}\frac{\partial U}{\partial x} - \frac{1}{g}\frac{\partial U}{\partial t} \right) \tag{2.6}$$

where $U$ is flow velocity, $x$ is the streamwise direction, and $\tau$ is time. Although we will not use this relation, an interpretation of it helps to illustrate one of the difficulties in estimating transport rates. To start, we note that if the flow were steady and uniform (meaning that all the derivatives in Eq. 2.6 equal zero), we recover our depth-slope product in Eq. 2.5. The first two terms after $S$ on the right side of Eq. 2.6 are the non-uniform flow terms, representing changes in the streamwise, or $x$, direction. The last term represents changes in time. The more rapidly the flow changes over $x$ (for example, flow through a bend, over a change in roughness or bed slope) or $t$, the larger will be the non-uniform and unsteady terms in Eq. 2.6.

The unsteady term ($\partial U / \partial t$) in Eq. 2.6 is typically important only with very rapidly changing flow, as with a dam break or surge. Dropping this term from Eq. 2.6, we get:

$$\tau_0 = \rho g R \left( S - \frac{\partial h}{\partial x} - \frac{U}{g} \frac{\partial U}{\partial x} \right) = \rho g R S_f \qquad (2.7)$$

where $S_f$ is the slope of the energy grade line—the imaginary surface connecting all points at an elevation representing the total mechanical energy in the flow—and is given by:

$$S_f = \frac{d}{dx} \left( z_b + h + \frac{U^2}{2g} \right) \qquad (2.8)$$

where $z_b$ is bed elevation and $U^2/2g$ is the velocity head ($S = -\partial z_b / \partial x$). $S_f$ is easily calculated in open channel flow models such as HEC-RAS (http://www.hec.usace.army.mil/software/hec-ras/).

In many cases, a flow model allowing computation of $S_f$ is unavailable and one is tempted to assume that the non-uniform flow terms are small, allowing use of Eq. 2.5 in determining $\tau_0$. You could *assume* that these derivative terms are small. This is sometimes true and sometimes incorrect. *How would you know?* If flow is changing rapidly (for example, due to a change in flow over time, through a constriction, or a change in slope or roughness), Eq. 2.6 indicates that the depth-slope product may produce a $\tau_0$ much different from the actual. Remember, small error in $\tau_0$ can produce large error in estimated transport rate. If the stage is known at several cross-sections for a specific discharge, values of the change in depth ($\Delta h$) and velocity ($\Delta U$) over the downstream distance ($\Delta x$) may be determined and used to estimate the magnitude of the terms in Eq. 2.7. If the estimated values of the non-uniform terms are much smaller than $S$, use of the depth-slope product is justified. This raises the very important distinction between an *approximation* (which can be evaluated quantitatively) and an *assumption* (which cannot).

So far, we have discussed how to estimate the total boundary stress $\tau_0$ in a stream reach. This gives us the *total* force acting on the wetted boundary of bed and banks. Some of this force acts on the movable grains on the stream bed and thus drives the transport, but some of it also acts on other things: woody and other debris in the channel, bridge piers, channel bends, and so forth. To estimate the sediment transport rate, we need to *partition* total stress $\tau_0$ into that part that acts only on the sediment grains. We'll call this the grain stress $\tau'$ (this is also called the skin friction). We have no direct way to estimate $\tau'$, although there are some useful approximate approaches. We will develop one approach here, based on the Manning Equation:

$$U = \frac{\sqrt{S}R^{2/3}}{n} \tag{2.9}$$

where $n$ is the Manning roughness. Eq. 2.9 is correct when $U$ and $R$ are expressed in m/s and m. If ft are used instead of m, then the right side of Eq. 2.9 must be multiplied by factor of 1.49. Typical values of $n$ for natural streams are in the range 0.025 to 0.08, although larger values are observed for very rough channels, particularly when they are clogged with vegetation.

A number of factors contribute to the boundary roughness and, therefore, to the magnitude of $n$. One source of roughness (the one we are interested in) is the bed grain size. You might reason (correctly) that larger grains would be hydraulically rougher than smaller grains. Using Eq. 2.9 this means that for the same $U$ and $S$, a bed with coarser sediment, and thus a larger $n$, will have a larger depth. An approximate relation between $n$ and a characteristic grain size of the bed material, often referred to as the Strickler relation, is:

$$n_D = 0.040D^{1/6} \tag{2.10}$$

for $D$ in m, or

$$n_D = 0.013D^{1/6} \tag{2.11}$$

for $D$ in mm. Figure 2.3 shows the variation of $n_D$ with $D$, along with the typical range of $n$ in gravel-bed rivers. The difference between the Manning-Strickler $n_D$ (given by Eqs. 2.10 or 2.11) and the actual $n$ indicates the effect of other factors increasing the bed roughness.

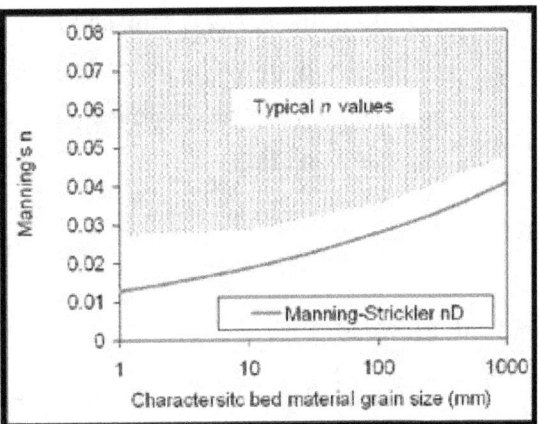

**Figure 2.3.** The Manning-Strickler $n$ relative to typical range of $n$.

Notice that Manning's equation contains both $R$ and $S$, suggesting we can solve it for $\tau_0$ via the depth-slope product (in fact, that is just what flow resistance equations are all about: a relation between velocity, flow geometry, boundary roughness, and $\tau_0$). If we multiply Eq. 2.9 by $(\rho g)^{2/3} S^{1/6}$ and rearrange, we get:

$$(\rho g)^{2/3} S^{1/6} nU = (\rho g RS)^{2/3} \tag{2.12}$$

Raising all this to the 3/2 power gives:

$$\rho g S^{1/4} (nU)^{3/2} = \tau_0 \tag{2.13}$$

Now, suppose we insert the Strickler definition of $n$ into Eq. 2.13. Recalling that other factors also contribute to $n$, the Manning-Strickler $n_D$ should be smaller than the total $n$ for the channel. By using the Manning-Strickler $n_D$ in Eq. 2.13, we are essentially calculating the shear stress due to the bed grains only, which is the approximation of $\tau'$ that we are after. Using Eq. 2.11 in Eq. 2.13, we get:

$$\rho g (0.013)^{3/2} (SD)^{1/4} U^{3/2} = \tau' \tag{2.14}$$

Now, we have to choose a grain size $D$ that represents the bed roughness. Hopefully, the larger sizes in the bed would tend to dominate the roughness. For example, $D_{90}$ and $D_{84}$ are often used because they are the grain sizes for which 90 percent or 84 percent of the bed material is finer. We will use $2D_{65}$, based on field and lab observations, although it is difficult to make a strong case for any particular value of $D$. Fortunately, the choice does not make a big difference because $D$ is found in Eq. 2.14 raised to the power ¼. Substituting $D=2D_{65}$ in Eq. 2.14 and using $r = 1000$ kg/m$^3$ and $g = 9.81$ m/s$^2$, we get:

$$\tau' = 17 (SD_{65})^{1/4} U^{3/2} \tag{2.15}$$

for $\tau'$ in Pa, $D_{65}$ in mm, and $U$ in m/s. We see that $\tau'$ depends mostly on the flow velocity (meaning that it depends on $Q$ and all the factors—channel size, shape, and slope—that determine flow depth and relate $Q$ and $U$) and, to a lesser extent, on $S$ and $D_{65}$.

# Transport Rate

## *Dimensional Analysis*

Bed-material transport rates are conveniently treated as a flux per unit width. We define transport rate per unit width, $q_s$, as the volume of sediment, $\forall_s$, transported per unit time and width $[L^2T^{-1}]$. To understand the constituents of a general transport model, it is useful to do a dimensional analysis. We can imagine that $q_s$ will depend on a number of variables representing the strength of the flow, fluid, and sediment. We use $\tau$ to represent the flow strength. We also include flow depth, h, in the list, arguing that interactions between the bed and water surfaces might alter the relation between $q_s$ and $\tau$ for shallower flows. We represent the sediment using grain size, $D$, and sediment density, $\rho_s$. Both of these control how heavy a grain is and $D$ also controls the grain area exposed to the flow and thereby the drag force acting on it. The balance between resistance to motion (which depends on grain weight) and flow force (which depends on grain area) should influence the transport rate. For now, we will pretend that the sediment contains only one size (a later section presents the difficult problem of representing grain size when you have a mixture of a wide range of sizes). We represent the fluid using water density, $\rho$, and water viscosity, $\mu$. Density, $\rho$, is the fluid mass per volume and governs the interaction between forces and accelerations in the fluid. For example, for the same $\tau$ and $D$, you can imagine that transport rates in air, which has very low density, would be different than transport rates in water). Viscosity $\mu$ describes the resistance of a fluid to deformation (for example, for the same $\tau$ and $D$, you can imagine that transport rates in a viscous motor oil would be different than transport rates in water or, more practically, that smaller grains with less mass might have a harder time moving through a viscous fluid than larger grains. Finally, we need to include the acceleration of gravity, $g$, which influences the movement of both the water and the sediment grains. Our list of variables is then:

$$q_s = f(\tau, h, D, \rho_s, \rho, \mu, g) \tag{2.16}$$

Our list has eight variables and these variables include the three fundamental dimensions of mass, length, and time. The rules of dimensional analysis tell us that we can reduce the list of eight variables by three (the number of fundamental dimensions), giving five dimensionless variables that represent all of the physical relations among the original eight variables. Although there are some strict rules governing dimensional analysis, there is no unique set of dimensionless variables that is the correct result of the analysis. Thus, there is some art and much practicality in the choice of dimensionless variables used. We do not present a complete dimensional analysis here, but accessible discussions can be found in Middleton and Southard (1984) and Middleton and Wilcock (1994). A common and useful set of dimensionless variables is:

$$q^* = f\left(\tau^*, S^*, s, D/h\right) \tag{2.17}$$

where

$$q^* = \frac{q_s}{\sqrt{(s-1)gD^3}}, \quad \tau^* = \frac{\tau}{(s-1)\rho gD}$$

$$S^* = \frac{\sqrt{(s-1)gD^3}}{\mu/\rho} \quad and \quad s = \frac{\rho_s}{\rho} \tag{2.18 a, b, c, d}$$

We have a dimensionless transport rate, $q^*$ (also known as the Einstein transport parameter), a dimensionless shear stress, $\tau^*$ (widely known as the Shields Number and sometimes given the symbol $\theta$), a dimensionless viscosity, $S^*$, relative grain density, $s$, and relative depth $D/h$. From the rules of dimensional analysis, we know that the relation among the five variables in Eq. 2.17 contains all the information in the relation among the eight variables in Eq. 2.16. If we are only concerned with quartz density grains in water (most sediment is close to quartz density, but we are excluding transport in air), we can drop $s$ from further consideration because it will be a constant. If we constrain ourselves to flow depths greater than a few times the grain size, $D$, we can argue that the relative flow depth, $D/h$, will have negligible effect. By this, we mean that the relation between $q^*$, $\tau^*$, and $S^*$ will not depend strongly on $D/h$. This will have to be confirmed with data and we can expect that the assumption might break down when shallow flows are diverted around, or tumbling over, coarse grains. Similarly, we know that if grains are coarser than one mm or so, the effects of viscosity on transport relations are relatively small, indicating that we might neglect $S^*$ for gravel transport.

Dimensional analysis has allowed us to identify two dimensionless variables governing transport rate and define conditions under which this short list of variables is likely to hold. For quartz density sediment coarser than about 1 mm, transported in water of depth more than a few times $D$, we propose that we can neglect the last three variables in Eq. 2.17, leaving only $q^*$ and $\tau^*$. Each has a nice physical interpretation. The transport variable, $q^*$, can be shown to represent the ratio of the volumetric transport rate, $q_s$, to the product ($wD$), where $w$ is the grain fall velocity. Thus, $q_s$ is scaled by the size and weight of the grain. The Shields Number, $\tau^*$, represents a ratio of the shear stress (flow force per area) acting on the bed to the grain weight per area.

### Transport Function for Uni-Size Sediment

Dropping $S^*$, $D/h$, and $s$ from the list in Eq. 2.17, we are left with:

$$q^* = f(\tau^*) \tag{2.19}$$

which says, in essence, that the rate of transport (relative to grain size and fall velocity) will depend on the flow shear force (relative to the grain weight). Transport functions often take a power form such as:

$$q^* = c\left(\tau^* - \tau_c^*\right)^d \tag{2.20}$$

where

$$\tau_c^* = \frac{\tau_c}{(s-1)\rho g D} \tag{2.21}$$

and $\tau_c$ is the critical value of $\tau$ necessary for initiating transport. The quantity $(\tau^* - \tau_c^*)$ is an expression for the "excess" shear above critical (another is $\tau^*/\tau_c^*$). For example, a well known empirical bed-load function is the Meyer-Peter and Müller (M-PM; Meyer-Peter and Müller 1948) formula:

$$q^* = 8\left(\tau^* - \tau_c^*\right)^{3/2} \tag{2.22}$$

Because it is quite simple and widely known, we will use M-PM to illustrate various aspects of sediment transport functions. Recent work (Wong and Parker 2006) suggests that the correct constant in M-PM should be 4 rather than 8. The actual choice of constant does not alter the principles we will illustrate and the use of M-PM in applications has been largely superceded by more recent formulas of somewhat different form, including the formulas implemented in BAGS.

In a later section, we will explain that the critical shear stress, $\tau$, is difficult to both define and measure for uni-size sediment and nearly impossible to measure for mixed-size sediments. For the purpose of estimating transport rates, it is both reasonable and useful to define a surrogate for $\tau_c$, the reference shear stress, $\tau_r$, which is the shear stress that produces a small, constant, and agreed-upon reference transport rate. By its definition, $\tau$ should be close to, but slightly larger than $\tau_c$. First, we define a new dimensionless transport parameter:

$$W^* = \frac{q^*}{\left(\tau^*\right)^{3/2}} = \frac{(s-1)g q_s}{\left(\tau/\rho\right)^{3/2}} \tag{2.23}$$

We use $W^*$ because it does not contain the grain size, $D$, which we will see later is an essential feature when developing a general model for the transport rates of sediments of different size or for different size fractions within the same mixture. The reference transport used is $W^* = 0.002$. For example, let's recast the M-PM formula using a reference transport rate. First, we divide Eq. 2.22 by $(\tau^*)^{3/2}$ to get:

$$W^* = 8\left(1 - \frac{\tau_c^*}{\tau^*}\right)^{3/2} \tag{2.24}$$

Now, we solve Eq. 2.24 for the reference value of $\tau^*$ (in other words, $\tau^* = \tau_r^*$ for $W^* = W_r^* = 0.002$). Dividing by 8 and raising both sides to the 2/3 power produces:

$$0.004 = 1 - \frac{\tau_c^*}{\tau_r^*} \tag{2.25}$$

from which we see that $\tau_c^*$ Thus, $\tau_r^*$ is slightly larger than $\tau_c^*$, as desired. Using this value to replace $\tau_c^*$ in Eq. 2.24, we get:

$$W^* = 8\left(1 - 0.996\frac{\tau_r^*}{\tau^*}\right)^{3/2} \tag{2.26}$$

which gives the M-PM formula in terms of $W^*$ and the reference shear stress.

## Transport Function for Mixed-Size Sediment

All gravel-bed rivers contain a range of sizes, so the work of the preceding section must somehow account for the range of sizes available for transport. The simplest approach is to assume that the function defined for uni-size sediment can be applied to a characteristic grain size for each mixture. In this case, the problem is to specify the characteristic grain size, for example, the median size $D_{50}$. This approach does not permit calculation of changes in transport grain size and, in fact, includes an implicit assumption that the transport grain size does not vary with transport rate, an assumption not consistent with observation.

The transport rate of individual size fractions, $q_{si}$, will depend on the grain size of each fraction $D_i$, and its proportion in the bed, $f_i$. A characteristic grain size for the overall mixture, $D_m$, is needed to determine the transport rate of the entire mixture and to define the relative size of fraction, $D_i$. Our list of dimensional variables is:

$$q_s = f(\tau, h, D_i, D_m, f_i, \rho_s, \rho, \mu, g) \tag{2.27}$$

Having added two variables to the list of dimensional variables for the uni-size case, we also add two to the list of dimensionless variables:

$$q* = f(\tau*, S*, s, D_m/h, D_i/D_m, f_i) \qquad (2.28)$$

The hypothesis used as the basis of many mixed-size transport models, including those in BAGS, is that the fractional transport rate, when scaled by the proportion of each fraction in the bed, will be a function of the Shields Number and critical Shields Number for each fraction:

$$q_i^* = f(\tau_i^*, \tau_{ci}^*) \qquad (2.29)$$

where

$$q_i^* = \frac{q_{si}}{f_i\sqrt{(s-1)gD_i^3}}, \tau_i^* = \frac{\tau}{(s-1)\rho g D_i}, \tau_{ci}^* = \frac{\tau_{ci}}{(s-1)\rho g D_i} \qquad (2.30 \text{ a, b, c})$$

The essential assumptions behind Eq. 2.29, to be tested against transport observations in developing the transport models, are:

(i) The proportion in each fraction, $f_i$, affects transport only as it determines how much of each fraction is available for transport. For example, changes in $f_i$ for one fraction are not assumed to influence the fractional transport rates of other fractions. Note that, for uni-size sediment, $f_i = 1$ and Eq. 2.30a reduces to Eq. 2.18a.

(ii) The effect on transport of $S*$, $s$, and $D_m/h$ are assumed to be negligible or contained in the critical Shields Number, $\tau_{ci}^*$.

(iii) The same functional relation in Eq. 2.29 holds for each size fraction in the mix.

The transport formulas in BAGS use the alternative dimensionless transport:

$$W_i^* = \frac{(s-1)gq_{si}}{f_i(\tau/\rho)^{3/2}} \qquad (2.31)$$

The absence of $D_i$ in $W_i^*$ facilitates the development of a transport function that holds for all sizes, as explained in the next section.

## How a Transport Model is Built

The transport models in BAGS were constructed using a similarity analysis. They begin with the hypothesis that the same transport function—a relation between $W_i^*$ and $\tau/\tau_{ri}$, where $\tau_{ri}$ is the reference shear stress for size fraction, *i*—applies to all fractions in all sediments. A "similarity collapse" is performed on the data, which means that all the transport data are plotted as $W_i^*$ versus $\tau/\tau_{ri}$ and the data are seen to collapse reasonably well about a common trend. The key feature of this pair of dimensionless variables is that neither contains grain size and, thus, the trends displayed by the data are not affected by the grain size of different fractions. We have not eliminated grain size from the problem, just from the transport function. In fact, what we have done is to isolate the influence of grain size (along with most other factors) to the reference shear stress, $\tau_{ri}$. Put another way, the similarity hypothesis states that, if we can determine $\tau_{ri}$ by whatever means, then we can predict the dimensionless transport rate $W_i^*$ using a single, general function of $\tau/\tau_{ri}$.

The process of building a transport model is clearer when illustrated with an example. Figure 2.4 shows fractional transport rates of a mixed sand-gravel sediment (part of the data used to produce the Wilcock-Crowe [2003] formula). Panel (a) shows $W_i^*$ as a function of $\tau$. The transport rates of the coarser fractions are considerably smaller than those of the finer fractions. The values of $\tau_{ri}$ selected for each size fraction are shown as "x"s on the reference transport line ($W^* = 0.002$). Panel (b) shows $W_i^*$ as a function of $\tau/\tau_{ri}$. Although scatter remains in the plot, the general trend of $W_i^*$ as a function of $\tau/\tau_{ri}$ is seen to be similar. More information on the similarity collapse used to develop transport models, including reference to earlier seminal work in Japan, can be found in Parker and others (1982) and Wilcock and Crowe (2003).

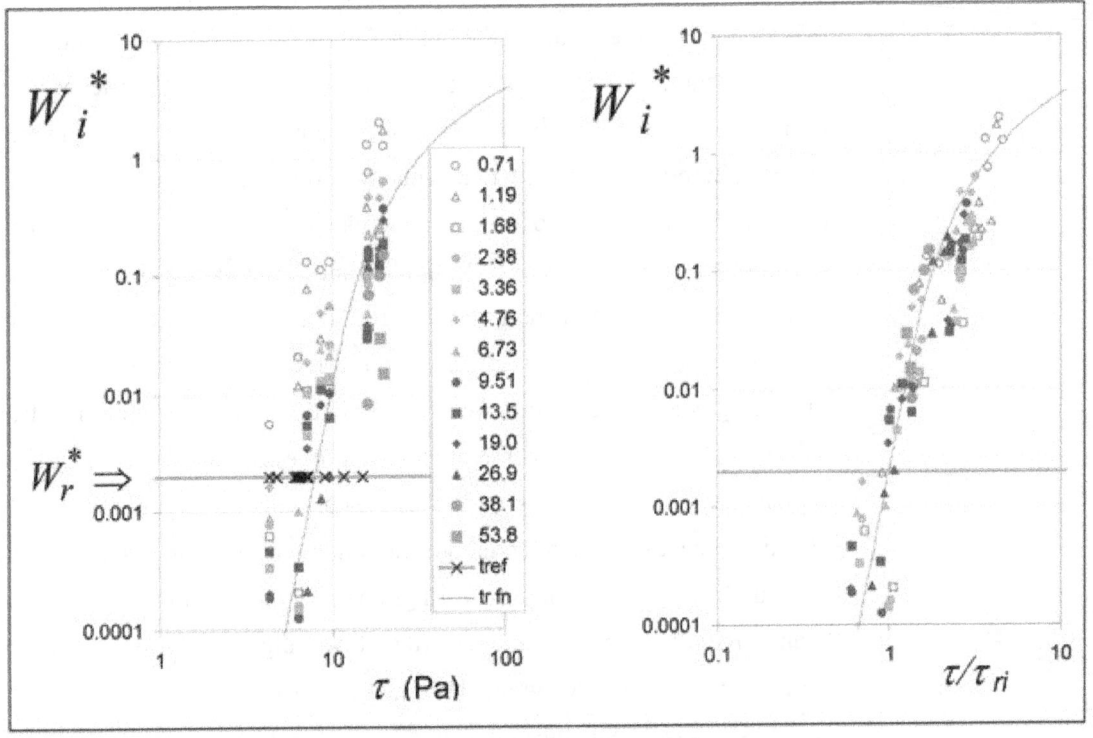

**Figure 2.4.** Illustration of the similarity collapse used to develop a transport model.

# Incipient Motion

## The Difference Between $\tau_c$ and $\tau_r$

So far, we have introduced the critical shear stress, $\tau_c$, and the reference shear stress, $\tau_r$. It can be easy to confuse them. The first, $\tau_c$, is well defined as an abstract concept—it is the value of $\tau$ at which transport begins. But because it is a boundary, it is impossible to measure directly. If you observe grains moving, then $\tau > \tau_c$. If no grains are moving, $\tau < \tau_c$. You could narrow this down with enough observations, but more difficult questions confound the issue. If you are looking for a grain to move, how long should you watch the bed and how much of the bed should you watch in order to determine whether grains are moving or not? When the flow is turbulent (meaning that $\tau$ at any point is fluctuating in time) and the size and configuration of the grains varies, these questions are difficult to answer. Yet, both are important (they affect the observed $\tau_c$) and detailed (Neill and Yalin 1969; Wilcock 1988). If our goal is to predict transport rate, the practical alternative is to use the reference shear stress, $\tau_r$, which is the value of $\tau$ associated with a very small, predetermined transport rate. This transport rate has been defined as $W^* = 0.002$ (Parker and others 1982). With measured transport rates over a range of small $\tau$, it is a straightforward thing to determine $\tau_r$. By its definition, $\tau_r$ is associated with a small amount of transport, so $\tau_r$ is slightly larger than $\tau_c$.

## Different Applications of Critical Shear Stress

Applications of the general concept of incipient motion can be divided into two broad categories. The first is that $\tau_c$ (or $\tau_r$) serves as an intercept, or threshold, in a sediment transport relation (as we have shown in the Meyer-Peter and Muller relation and illustrated in fig. 2.4). The presence of $\tau_c$ (or $\tau_r$) in transport relations introduces a characteristic concave-down trend to the transport function (fig. 2.4). For the purpose of estimating transport rates, we are not concerned with the entrainment of any grain in particular, but need to know the flow associated with some particular transport rate. The reference shear stress, $\tau_r$, was developed for this purpose. In the second case, we *are* interested in the entrainment of individual grains. For example, we might be interested in flushing fines from the substrate of a gravel-bed river in order to improve spawning habitat. Or we might be interested in the stability of bed and bank material in cases where channel stability depends on the material not moving at all. In these cases, we are interested in the entrainment of individual grains or, more generally, the proportion of grains on the bed surface that are entrained. We might ask "At what discharge do 90 percent of the surface grains become entrained, thereby providing access to the substrate and some flushing action?" On the other hand, "At what discharge do 1 percent of the surface grains become entrained, thereby indicating that our rip-rap channel is beginning to fall apart?"

The difference between these two applications of incipient motion can be illustrated with their characteristic field methods. As an intercept in a transport relation, we would determine $\tau_r$ by measuring transport rate and determining the value of $\tau$ at which the transport rate is equal to a small reference value. In contrast, the simplest way to measure actual bed entrainment is to use tracer grains. These might be painted rocks that are placed on the bed surface (generally, we try to replace an *in situ* grain with a painted grain of the same size in order to provide a more realistic indication of the flow producing movement). If the streambed (or a portion of it) is dry, it is even easier to just spray paint the bed itself, although this may raise aesthetic or legal objections. After a flow has passed over the bed, the number of painted rocks remaining are counted. Tracers provide an excellent (and easy) way of measuring entrainment (Did the grains move at all?), but it is difficult to determine transport rates from tracers, which requires relocating a large fraction of the tracers and determining how far they moved. Entrainment of 50 percent of the grains on the bed does not tell you what the transport rates were. And measurement of a non-zero transport rate does not tell you how many of the surface grains were entrained. A significant transport rate could be produced by a few hyperactive grains, while most of the grains on the bed surface don't move at all.

A related concept is *partial transport*, which is defined as the condition in which only a portion of the grains on the bed surface ever move over the duration of a transport event. We could define partial transport in terms of all surface grains (for example, 50 percent of the surface grains move over the transport event) or on a size-by-size basis (for example, 90 percent of the 2- to 8-mm grains move, 50 percent of the 8- to 32-mm grains move, and only 5 percent of the >32-mm grains move over the transport event). The scope and nature of partial transport was defined in the laboratory (Wilcock and McArdell 1997) and has been shown to represent transport conditions in the field, even under large flow events (Haschenburger and Wilcock 2003; Hassan and Church 2000). Beyond its importance in terms of defining bed stability and substrate flushing, partial transport appears to have important consequences for defining frequency and intensity of benthic disturbance in the aquatic ecosystem.

## Incipient Motion of Uni-Size Sediment

The dimensional analysis for uni-size sediment transport rate led to the result that dimensionless transport rate depended on four dimensionless variables, the Shields Number, $\tau^*$, a dimensionless viscosity, $S^*$, the relative density, $s$, and the relative flow depth, $D/h$. If we argue that the variables that determine transport rate are the same as those that determine whether grains are moving or not, then the same dimensional analysis also applies to incipient motion if we simply replace $q_S$ with a "motion/no motion" binary variable. Incipient grain motion should be described by some relation between $\tau_c^*$, $S^*$, $s$, and $D/h$. If, as we did before, we limit ourselves to typical values of $s$ (2.65±5 percent) and flow depths more than a few times $D$, we end up with a relation between $\tau_c^*$ and $S^*$. For uni-size sediments, this is represented by the widely known Shields diagram.

The trend marked Shields on the diagram is the function

$$\tau_c^* = 0.105\,(S^*)^{-0.3} + 0.045 \exp\left[-35\,(S^*)^{-0.59}\right] \qquad (2.32)$$

which approximates the original Shields curve (as amended by Miller and others 1977) and allows $\tau_c^*$ to be determined without having to look values up on the diagram. The curve marked "Surface" on figure 2.5 is the function:

$$\tau_c^* = \frac{1}{2}\left[0.22\,(S^*)^{-0.6} + 0.06 \cdot 10^{[-7.7(S^*)^{-0.6}]}\right] \qquad (2.33)$$

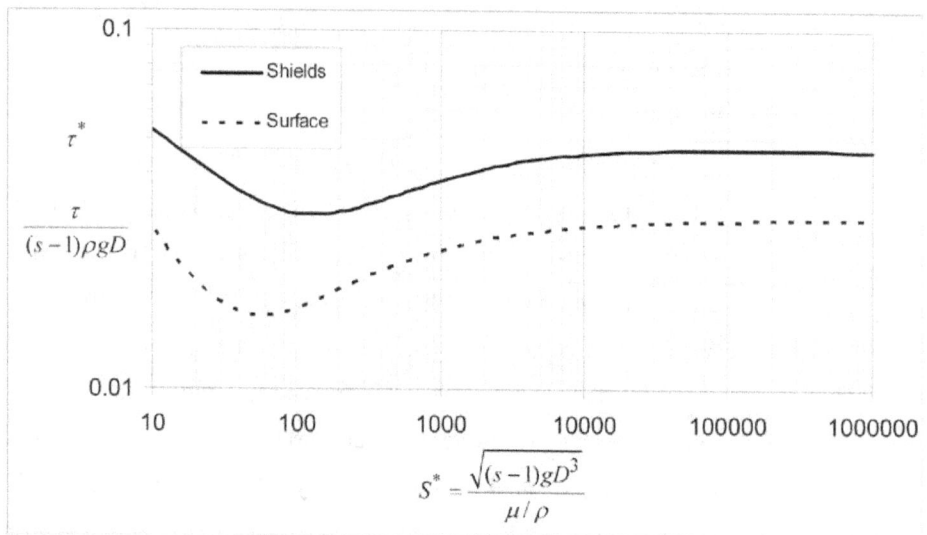

**Figure 2.5.** Shields diagram for incipient motion of uni-size sediment.

This is a function fitted to the Shields Curve by Brownlie (1981), but multiplied by 0.5, which Parker and others (2008) proposed to match Neill's (1968) observation that $\tau_c^* = 0.03$ at large $S^*$. This suggests that this "surface" curve is more appropriate for estimating $\tau_c^*$ when a pebble count is used to measure the grain size of the bed surface.

The variation of $\tau_c^*$ with $S^*$ demonstrates the effect of fluid viscosity on grain entrainment. Grains smaller than a few mm are associated with $S^*$ of order

1000. At smaller $S^*$, we see that $\tau_c^*$ varies with $S^*$, indicating that viscosity influences $\tau_c^*$ for smaller grains. For coarser grains, $\tau_c^*$ approaches a constant value of about 0.03. This is of particular interest, because we are interested in gravel-bedded streams. Using the definition of $\tau_c^*$, we see that $\tau_c^* = 0.03$ corresponds to:

$$\tau_c = 0.03(s-1)\rho g D \tag{2.34}$$

and using $s = 2.65$ and $rg = 9810$ kg m$^{-2}$ s$^{-2}$ we get:

$$\tau_c = 0.5D \tag{2.35}$$

for $\tau_c$ in Pa and $D$ in mm. This linear trend is clear when the Shields diagram is plotted for $\tau_c$ in Pa and $D$ in mm (fig. 2.6).

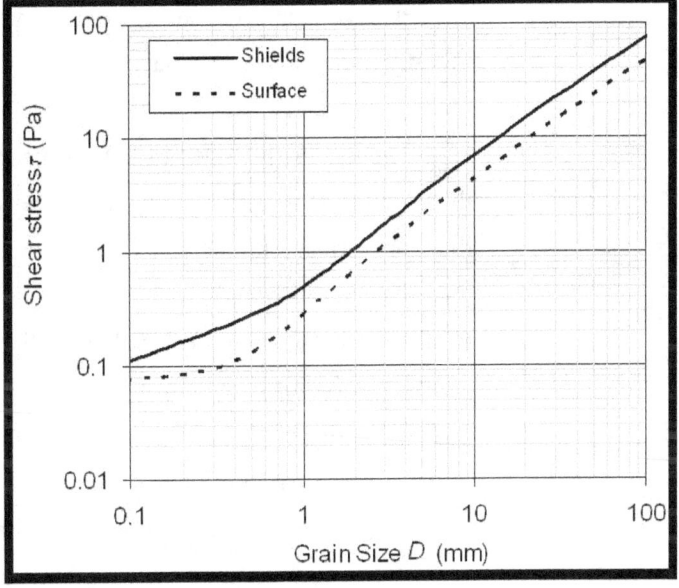

**Figure 2.6.** Shields Curve in dimensional space.

## Incipient Motion of Mixed-Size Sediment

What if the bed material contains a range of sizes? The tendency for larger grains to be harder to move (as reflected in Eq. 2.35 and the dimensional Shields curve above) is counterbalanced by the effect of mixing the different sizes together in the same sediment. When placed in a mixture, smaller grains will be harder to move than when in a uni-size bed and larger grains will be easier to move. For sediments that are not too widely sorted with a unimodal size distribution (for example, a mixture of medium sand to pea gravel with $D_{50}$ around 2 mm, or a gravel in the range 8 mm to 64 mm with a prominent mode in the 21 mm to 32 mm fraction and little sand), it turns out that all of the different sizes in the mixture have just about the same value of $\tau_c$. This is a key element of the condition of "equal mobility," which was hotly debated 10 to 20 years ago (Komar 1987; Parker and others 1982). For equal mobility, the balance between the *absolute size* effect (as in the Shields Diagram) and the *relative size effect* is exact, meaning that all sizes begin moving at the same stress. This is the basis of the Parker-Klingeman-McLean model included in BAGS. The transport rate of all sizes (relative to their proportion in the bed) is the same. The transport has the same grain size as the bed and can be predicted using the transport rate of a single size.

If all sizes have the same $\tau_c$, what is it? If the grain size standard deviation is not too large, it turns out that the Shields curve provides a pretty good indication of $\tau_c$ if the median grain size $D_{50}$ is used for $D$ (Wilcock 1993). Thus, you can approximate $\tau_c$ for the entire mixture, as well as for individual size fractions, using the Shields Diagram and $D = D_{50}$.

For mixtures with a wider range of sizes, the smaller sizes tend to move at smaller flows than the coarser sizes. This is particularly the case with bimodal mixtures, which typically have a primary mode in the gravels and a secondary mode in the sand sizes.

The models included in BAGS incorporate different functions to describe the variation of $\tau_{ri}$ with relative grain size $D_i/D_{50}$ or $D_i/D_m$ (fig. 2.7). Collectively, these are termed "hiding functions," a name suggesting that mixture effects reduce the mobility of only smaller grains, although the hiding functions are applied to all sizes and thus incorporate both hiding of fine grains and exposure of coarse grains. The hiding function of the Parker-Klingeman model gives $\tau_{ri}$ as a power function of $D_i/D_{sub50}$, where $D_{sub50}$ is for the substrate size distribution. The Parker surface-based model gives $\tau_{ri}$ as a power function of $D_i/D_{sg}$, where $D_{sg}$ is for the surface size distribution with the sand excluded. The Wilcock/Crowe surface-based model uses a hiding function that varies with relative grain size, $D_i/D_{sg}$, in a more complex fashion. The difference between the hiding functions arises in large part because the Wilcock/Crowe model is based on an extensive set of coupled observations of transport and surface grain size. It indicates that the fractions finer than $D_{sg}$ are nearly equally mobile whereas the coarser fractions display considerable variation of $\tau_{ri}$ with $D_i/D_{sg}$. Further discussion of these transport models is given in Chapter 4.

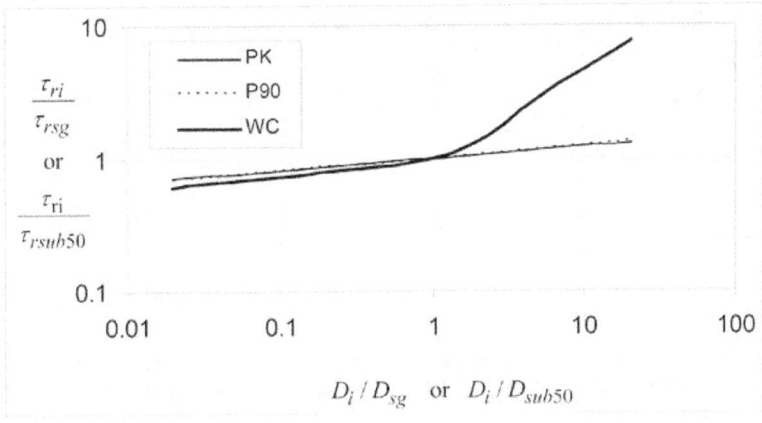

**Figure 2.7.** Hiding functions, reference shear stress as a function of relative grain size, used in the Parker-Klingeman, Parker, and Wilcock/Crowe transport models.

# The Effect of Sand and a Two-Fraction Transport Model

Examination of the reference shear stress for a wide range of sediments indicates that the range of $\tau_{ri}$ is relatively small for many sediments (as suggested by the PK and P90 hiding functions in fig. 2.7), but that some sediments show a much larger variation (as suggested by the WC hiding function in fig. 2.7). The sediments with little variation in $\tau_{ri}$ have size distributions that are unimodel in shape and contain little sand. Those with a broader range of $\tau_{ri}$ are bimodal, with one mode in the gravel size range and another mode in the sand size range. The reference shear stress, $\tau_{ri}$, for the sand fractions tends to be much smaller than for the gravel fractions (in others words, the sand begins moving at smaller flows than the gravel, violating the equal mobility condition) and $\tau_{ri}$ for the gravel fractions tends to be smaller in sandy mixtures than in mixtures with little sand (Ikeda and Iseya 1988). The sand content affects $\tau_{ri}$ for *both* the sand and gravel fractions.

The bimodal nature of the size distributions and previous observations that $\tau_{ri}$ does not vary much within unimodal gravel mixtures suggests that differences between the mixtures might be resolved by considering the mixtures as being composed of two fractions—sand and gravel. This allows each fraction to have a different $\tau_{ri}$ and provides a simple basis for representing the effect of sand content on $\tau_{ri}$. It turns out that such a two-fraction approach for describing the bed material size distribution captures these points very well. This is shown by the plots of the reference Shields number of the gravel $\tau_{ri}^*$ and sand $\tau_{rs}^*$ fractions for five different lab sediments and four different field cases as a function of the proportion of sand in the bed $f_s$ (fig. 2.8; Wilcock 1998; Wilcock and Kenworthy 2002).

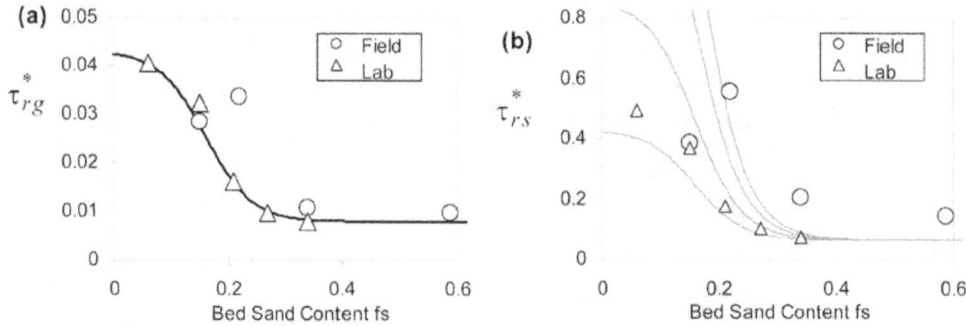

**Figure 2.8.** Variation of reference Shields Number for (a) gravel and (b) sand fractions of five laboratory sediments and four field cases. The curves in (b) are for $D_{gr}/D_s$ = 10, 20, 35, and 50.

The trends shown in figure 2.8 follow a pattern that fits nicely with our general understanding of transport. For the gravel, $\tau^*_{rg}$ approaches a common uni-size value of 0.045 as the sand content goes to zero. As sand content becomes large, $\tau^*_{rg}$ approaches a minimum of about 0.01, a value observed in different kinds of lab experiments. Most striking is the decrease in $\tau^*_{rg}$ over a range in sand content between about 10 percent and 30 percent. Over this range, the bed under-goes a transition from being *framework supported* (meaning that the bed consists of a framework of gravel clasts) to being *matrix supported* (meaning that the coarse grains are "floating" in a matrix of sand). This change in bed composition is clearly related to an associated change in transport behavior.

The trend in $\tau^*_{rg}$ is also clear, but more complex. As sand content approaches 1, $\tau^*_{rs}$ approaches a standard uni-size value, just as for the gravel. As sand content approaches zero, we can expect this small amount of sand to settle down among the gravel grains and sand entrainment to occur only when the gravel moves—thus, $\tau_{rs} = \tau^*_{rg}$. By the definitions of $\tau^*_{rs}$ and $\tau^*_{rg}$, $\tau^*_{rs} = \tau^*_{rg}(D_{gr}/D_s)$ (where $D_{gr}$ is the median size of the gravel fractions and $D_{gr}$ is the median size of the sand frac-tions), so $\tau^*_{rs}$ will depend not only on $\tau^*_{rg}$, but on $(D_{gr}/D_s)$. The multiple lines in the lower diagram are for different values of $(D_{gr}/D_s)$.

Recalling that the relation between transport rate and $\tau$ is very steep and nonlinear over the typical range observed in gravel-bed rivers, the magnitude of the shift in critical Shields Number shown in figure 2.8 is striking. Very large differences in transport rate can occur with a shift in sand content (Curran and Wilcock 2005; Wilcock and others 2001).

A two-fraction approach to modeling sediment transport, as suggested by the difference in behavior between fine and coarse bed material load, provides an approach that has both conceptual and practical advantages. Its conceptual under-pinnings derive from the essential simplification of equal mobility, revised to state that the sizes *within* two separate fractions—sand and gravel—are equally mobile even though the fractions themselves differ in their mobility. A two-fraction estimate allows sand and gravel to move at different rates, thereby permitting change in bed grain size due to changes in the relative proportion of sand and gravel (if not due to the changes in the representative grain size of either fraction). This provides a means of predicting the variation in the fines content of the bed, which may often be more variable than that of the coarse fraction, and whose passage, intrusion, or removal may be a specific environmental or engineering objective.

A two-fraction approach provides a ready means of representing the interaction between the fine and coarse components of the bed material. Laboratory studies (Curran and Wilcock 2005; Wilcock and others 2001) show that the addition of sand to a gravel bed or to the sediment supply can increase gravel transport rates by orders of magnitude (this is indicated by the four-fold decrease in $\tau_{rg}^{*}$ in fig. 2.8a). Because there are a variety of situations in which the supply of fine bed material can be increased (for example, fire, reservoir flushing, dam removal, urbanization), an accurate and practical basis for addressing these situations is needed.

A two-fraction approach to modeling sediment transport has a practical advantage in that it facilitates developing an estimate of the grain-size of an entire river reach. Areas with similar fines content may be mapped and combined to give a weighted average proportion of sand for the reach giving an integral measure of grain size with reasonable effort. This provides a superior description of the bed compared to an unsupported extrapolation from detailed sampling at only a few locations. Discussion of concepts and methods for mapping bed texture can be found in Buffington and Montgomery (1999a) and Bunte and Abt (2001).

BAGS supports a two-fraction computation of transport using a transport formula (Wilcock and Kenworthy 2002) and a calibrated approach (Wilcock 2001). Details are provided in Chapter 4.

# Chapter 3—Sources of Error in Transport Modeling

## It's the Transport Function

The form of a transport function reveals why transport rates are so difficult to estimate. We will use the Meyer-Peter and Müller formula (fig. 3.1) to illustrate (other transport formulas have a similar shape, so we use the simple and familiar M-PM formula as an example). Although the power function approaches an exponent of 3/2 at large $\tau^*$, it also becomes arbitrarily steep as $\tau^*$ approaches $\tau_c^*$ (fig. 3.1a). Small errors in $\tau^*$ near $\tau_c^*$ will produce order-of-magnitude errors in $q^*$. Unfortunately, most transport in coarse-bedded rivers occurs at $\tau^*$ not much larger than $\tau_c^*$ (even during floods, $\tau^*/\tau_c^*$ often does not exceed two; Parker and Klingeman 1982). Figure 3.1b shows the same relation using arithmetic axes. It is seen that the transport rates are zero and then increase rapidly as $\tau^*$ exceeds $\tau_c^*$. What this means is that small errors in $\tau^*$ (or $\tau_c^*$) can produce enormous errors in estimated transport rate. Recalling the definition of $\tau^*$

$$\tau^* = \frac{\tau}{(s-1)\rho g D} \qquad (3.1)$$

reminds us that $\tau^*$ contains $\tau$ in the numerator and $D$ in the denominator. Thus, we can summarize the problem of accurately estimating transport rate in terms of uncertainty in estimating $\tau$, which we will call the flow problem, and uncertainty in estimating $D$, which we will call the sediment problem. A third problem concerns uncertainty in specifying the correct value of $\tau_c^*$. In any of these cases, we see that an error of order, say, a factor of two, can produce large errors in $q^*$.

To illustrate the error possible, the same calculations are plotted in more familiar units in figure 3.2, which gives transport rate in kg/hr as a function of discharge in m³/s using M-PM with $\tau_c^* = 0.045$ and applied to a rectangular channel of width $b = 15$ m, slope $S = 0.002$, and roughness $n = 0.025$. The effect of uncertainty in the key variables ($\tau$, $D$, $\tau_c^*$) is evident not only in the steepness of the curves, but by using two grain sizes, $D = 45$ mm and $D = 30$ mm. The critical discharge for incipient motion differs by almost a factor of two between the two cases. At a discharge of 55 m³/s, M-PM predicts zero transport for 45-mm grains and a transport rate of 80,000 kg/hr for 32-mm grains. The identical two curves would be calculated for $D = 45$ mm and two values of critical Shields Number:

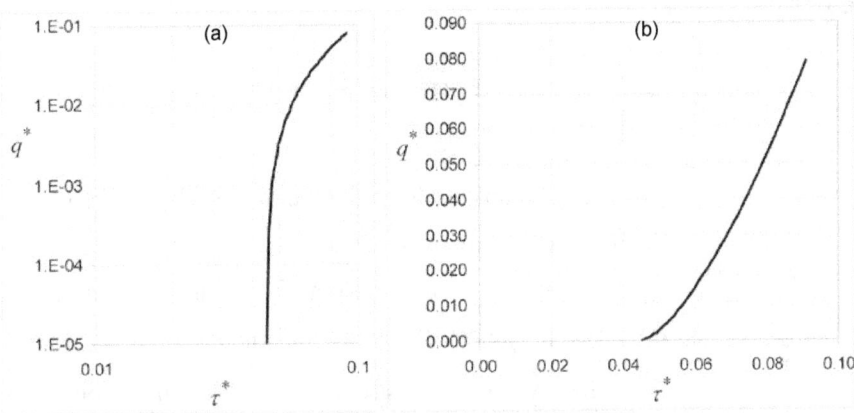

**Figure 3.1.** The Meyer-Peter and Müller formula, using $\tau^*_c = 0.045$. Formula shown with both log-log and arithmetic axes to demonstrate its nonlinearity. The log-log plot emphasizes small transport rates and demonstrates the zero limit at $\tau^*_c = 0.045$. The arithmetic plot shows how transport rate increases rapidly as $\tau^*$ exceeds $\tau^*_c = 0.045$.

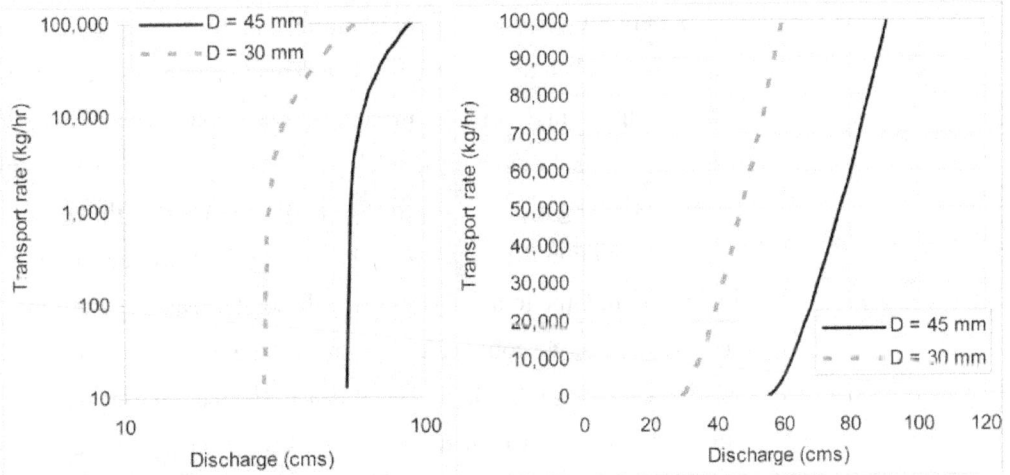

**Figure 3.2.** Example calculation using the Meyer-Peter and Müller formula for a channel with slope $S = 0.002$, roughness $n = 0.025$, and width $b = 15$ m. The solid curve uses $\tau^*_c = 0.045$ and $D = 45$ mm. The dashed curve can be generated using either $\tau^*_c = 0.045$ and $D = 30$ mm or $\tau^*_c = 0.030$ and $D = 45$ mm. Note that at discharge $Q = 55$ m$^3$/s, one curve indicates zero transport and the other a transport rate of 80,000 kg/hr.

$\tau^*_c = 0.045$ and $\tau^*_c = 0.030$. The same discrepancy would also arise from an error in $\tau$ of the same magnitude. Considering the uncertainty involved in specifying values of $\tau$, $D$, or $\tau^*_c$, you see that *it is very easy to be very far off with transport calculations.*

In this chapter, we summarize the three sources of error: error in $\tau$ (the flow problem), error in $D$ (the sediment problem), and error in $\tau^*_c$. It is quite easy to have error of a factor of two or larger in any and all of these terms. In Chapter 7, we will examine how to estimate error in transport rates given the possible combined error in all of these variables. Here, we explore each source of error individually to understand not only the error, but possible means of reducing it.

# The Flow Problem

There are three basic reasons why it is difficult to estimate the $\tau$ driving the transport:

1. Unsteady and non-uniform flow: although often neglected, accelerations in the flow in space or time can have a substantial effect on the total stress $\tau_0$ acting on the wetted perimeter of the channel. The components of $\tau_0$ were discussed in Chapter 2—Non-Uniform and Unsteady Flow.

2. Total stress versus grain stress, or skin friction: although we can estimate the total force per area acting on the channel boundary, only a portion of this total stress acts on movable grains to produce transport. The stress that acts on the movable grains $\tau'$ is called the grain stress. Methods for estimating $\tau'$ are approximate and were discussed in Chapter 2—The Drag Partition.

3. Spatial variability: $\tau$ tends to vary across and along the channel. The interaction between stress and grain motion is largely played out at a scale of perhaps 10 to 100 times the grain size. This dynamic is largely unaffected by what is happening elsewhere in the channel—it is a *local* phenomenon. In a channel with bed topography, the local shear stress $\tau_l$ can vary considerably across and along a stream reach (fig. 1.1). Combined with the fact that the transport function is steep and nonlinear (fig. 3.2), we face a difficult problem in determining the total transport rate through a reach. If the transport function was linear, we could calculate an average value of $\tau'$ and then use that in our transport formula. Because the transport function is strongly nonlinear, this will produce errors that can be significant. Basically, the surplus in transport rate in areas where $\tau'$ is greater than the mean will be much larger than the deficit in transport rate in areas where $\tau'$ is smaller than the mean. A simple case would be one in which the mean $\tau'$ is less than $\tau_c$, indicating that no transport should occur. But even if the mean $\tau' < \tau_c$, there can still be locations where $\tau_l > \tau_c$. Thus, the mean $\tau'$ indicates no transport when, in fact, there will be transport going on.

There are ways to estimate local shear stress, but these require local measurements or extensive detailed information about the channel topography and bed material (Wilcock 1996). In a research context, both $\tau_l$ and $q_s$ might be measured across a channel section. A practical interest in $\tau_l$ arises, for example, if one were interested in the spatial distribution of transport or the likelihood of entrainment over particular locations on the bed, such as over salmonid redds or dynamic gravel bars. These questions of detail are beyond the scope of our work here. There is some basis for estimating the increase in total transport through a section relative to the mean as a function of cross-section topography (Paola and others 1999), but application requires detailed measurement of section topography. Some transport formulas make a simple adjustment for transport in the field relative to that in flumes (where topographic relief is smaller; Brownlie 1981; Wilcock and Kenworthy 2002), although the adjustment can only be approximate because the true correction depends on the topography of any particular reach.

## The Sediment Problem

### Determining Grain Size

Grain size enters sediment transport calculations in two ways, via the critical stress $\tau_c$ or reference stress $\tau_r$ and via the proportion $f_i$ of each fraction in the bed if the transport of individual sizes is to be calculated. For coarse sediment, $\tau_r$ depends approximately linearly on grain size. Thus, error in specifying grain size produces a similar error in $\tau_r$. Because the transport rate depends nonlinearly on the excess $\tau'$ above $\tau_r$, the corresponding error in transport rate is likely to be much larger. Fractional transport rates depend in a linear fashion on $f_i$ via Eq. 2.31, although the proportion of sand in the bed also affects the reference shear stress in the Wilcock/Kenworthy two fraction model and the Wilcock/Crowe many-fraction model.

### Which Grain Size: Supply or Bed?

When determining the grain size to use in estimating transport through a reach, the obvious location to measure the grain size would be in the reach itself. But, the grain size of sediment transported through a reach may be considerably different from that found in the bed of the reach. An extreme case would be transport through an armored reach below a dam. If the reservoir traps all upstream sediment supply and flows capable of moving sediment are still released from the dam, the transportable sediment will be gradually removed from the reach downstream (Williams and Wolman 1984). If a tributary below the dam supplies sediment to the mainstem, the transported material will be that supplied by the tributary, which could be very different from that remaining in the armored river bed. For example, if the armored bed is coarse gravel and cobble and the tributary contributes sand and fine gravel at a rate that can be readily transported by flows released from the dam, then the mainstem bed will contain little or no sand and fine gravel and the grain size of the transport will be entirely different from the grain size of the bed. A transport formula applied to the grain size of the bed would produce an estimate of negligible transport, when, in fact, substantial transport of tributary sediment may be occurring.

Although transport in an armored reach below a dam may seem an extreme (but not uncommon) case, it is always possible that transport grain size is different from that of the bed (Lisle 1995) and different from that which would be predicted from the grain size of the bed. In coarse-bedded streams, it is simply not possible to reliably estimate transport rates by using the bed material grain size in a transport formula. Some indication of the grain size of the active portions of the streambed and/or the transported sediment is needed. Look for locations with fresh deposits to provide a check on the estimated transport grain sizes. The best approach is to use some samples of transport rate to calibrate the prediction from a transport formula, as developed in the Wilcock two-fraction and the Bakke and others many-fraction methods discussed in the next Chapter.

## The Incipient Motion Problem

A wide variation has been reported in $\tau_c^*$ (Buffington and Montgomery 1997). Some of this variation is due to intrinsic variability in the stress needed to entrain grains of a given size, but some of the variation may also be attributed to the flow and sediment problems discussed above. If $\tau_c^*$ is determined for a river reach, uncertainties due to non-uniform flow, drag partitioning, and local flow variation will contribute to uncertainty in $\tau_c^*$. There is certainly variation in $\tau_c^*$ intrinsic to the sediment. A factor of two variation in $\tau_c^*$ is reported due to the arrangement of gravel clasts into clast jams, stone lines, and stone cells on the stream bed (Hassan and Church 2000). A factor of four variation in gravel $\tau_c^*$ has been demonstrated as a function of sand content (fig. 2.8). Long periods without active transport can also increase $\tau_c^*$ from bed consolidation, chemical precipitation, and deposition of muds and biofilms.

## Use of Calibration to Increase Accuracy

Using a few transport measurements increases the accuracy of transport estimates by addressing each of the problems presented so far in this chapter. In both the Wilcock two-fraction method and the Bakke and others many-fraction method, the transport observations are used to determine the reference shear stress (for sand and gravel in the first case, and for each size fraction in the second case). This can be visualized by placing the transport data and the transport function on the same plot. Figure 3.3a shows some sample transport data and the Wilcock two-fraction transport function as $W^*$ plotted against $\tau'$. The same data and function are plotted in figure 3.3b except that $\tau'$ is now scaled by $\tau_r$. The calibration consists only of choosing a value of $\tau_r$ such that the function passes through the data.

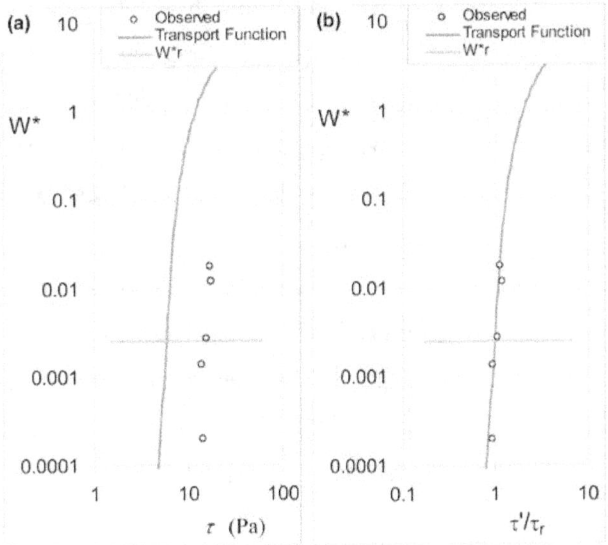

**Figure 3.3.** Illustration of transport formula calibration via $\tau_r$.

The calibration evidently accounts for uncertainty in the critical Shields Number (Chapter 3—The Sediment Problem) by finding $\tau_r$ from the data. But it accomplishes more than that because it can be used to, in effect, perform the drag partition. To see this, first recall the drag partition formula:

$$\tau' = 17(SD_{65})^{1/4}U^{3/2} \qquad (2.15)$$

Now, we assume that the variation of $U$ with $Q$ is known from a calibrated stage-discharge or flow resistance relation. To illustrate, consider a velocity-discharge rating curve represented with a power function (as used in at-a-station hydraulic geometry):

$$U = kQ^m \qquad (3.2)$$

where $k$ and $m$ are known from measurements. Putting Eq. 3.2 in Eq. 2.15, we have at the reference transport condition:

$$\tau'_r = 17k(SD_{65})^{1/4}Q_{ri}^{3/2m} \qquad (3.3)$$

If we write this equation a second time, for the general case of $\tau'$ as a function of $Q$:

$$\tau' = 17k(SD_{65})^{1/4}Q_{ri}^{3/2m} \qquad (3.4)$$

and then divide Eq. 3.4 by Eq. 3.3, we get:

$$\frac{\tau'}{\tau'_r} = \left(\frac{Q}{Q_r}\right)^{3/2m} \qquad (3.5)$$

if we can assume that $S$ and $D_{65}$ do not vary with $Q$. Now we see how calibration provides the drag partition. By providing, through calibration, an accurate estimate of $\tau_{ri}$ (and $Q_{ri}$ via Eq. 3.3), we now only need to predict the variation of $\tau'$ with $Q$ rather than the particular value of $\tau'$ at any particular $Q$. This variation depends only on $m$, which standard field methods can provide with reasonable accuracy.

Because the grain size of the transport samples is measured, the calibration procedure can be performed for each grain size separately. Using calibration and Eq. 3.5, it is seen that the predicted transport rates do not depend explicitly on grain size. It is no longer necessary to put an estimate of the grain size into the transport formula.

# Chapter 4—Transport Models in BAGS

This chapter describes the particular transport models implemented in BAGS, along with a few general comments on their application.

1. The substrate-based equation of Parker-Klingeman-McLean (PKM) (Parker and others 1982)

2. The substrate-based equation of Parker-Klingeman (PK) (Parker and Klingeman 1982)

3. The surface-based equation of Parker (P90) (1990a,b)

4. The procedure of Bakke and others (1999)

5. The two-fraction equation of Wilcock (W01) (2001)

6. The surface-based equation of Wilcock and Crowe (WC) (2003)

All of these relations are empirical—all were developed and tested using gravel or sandy gravel transport data. Four of the formulas are based directly or indirectly on the transport measurements of Milhous (1973) on Oak Creek, a gravel-bed stream in the Coast Range of Oregon. Three of the formulas (Parker 1990a,b; Parker and others 1982; Parker and Klingeman 1982) were developed directly from these data and have subsequently been applied to a large number of different gravel-bed rivers. The Bakke and others (1999) method is a calibrated procedure based on the Parker-Klingeman formula. The two-fraction model of Wilcock (2001) was developed from a mix of laboratory and field observations, including the Oak Creek data. The Wilcock-Crowe model was developed from a new set of flume data, described in Wilcock and others (2001).

Although these relations are often referred to as bed-load equations, they are, with the exception of the Parker surface-based transport model, bed-material load equations because they are fitted to the entire load. Although at low flows most or all of the grains may move solely as bed load, at higher flows the smaller fractions may move via a complex and difficult-to-distinguish mix of bed load and suspended load.

## General Comparison of the Transport Models

There is little substantial difference in the transport function forming the backbone of each of these transport models. Figure 4.1 presents all five transport functions (the Bakke procedure is a method of calibrating the Parker-Klingeman formula) contained in BAGS. The PKM, PK, and W01 formula for gravel are seen to be nearly identical. The two surface-based transport functions (P90 and WC) and the W01 sand function have a somewhat gentler slope and smaller values, which reflect a difference between using surface and substrate grain sizes and, to some extent, the effect of surface sand content on gravel transport rates. The form of the P90 transport function depends on the standard deviation of the surface size distribution. The curve plotted in figure 4.1 uses $\sigma_\psi = 1.5$, where $\sigma_\psi$ is the standard deviation of the grain size distribution in $\psi$ units. The Meyer-Peter and Müller relation is shown on figure 4.1 for reference. It has a larger slope at small $\tau/\tau_r$ and sharper inflexion than the newer relations.

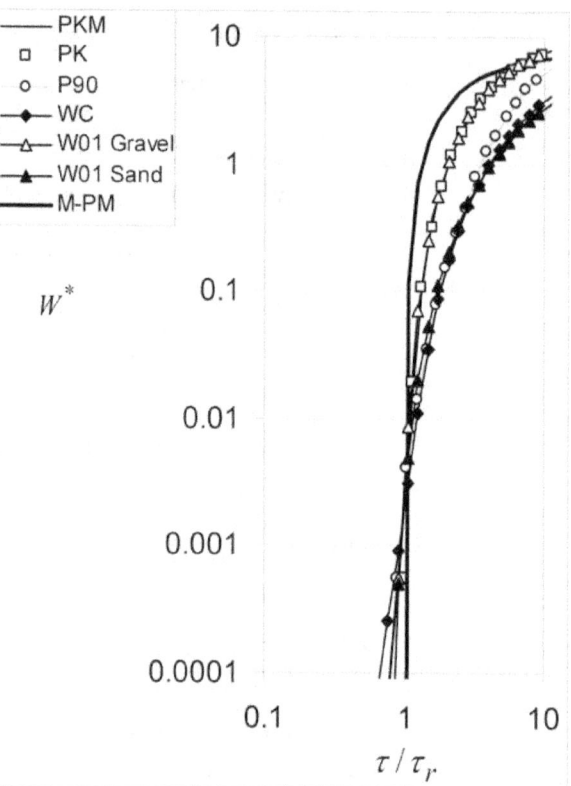

**Figure 4.1.** Transport function for each model ($W^*$ v. $\tau/\tau_r$).

The similarity among the transport functions is to be expected because the transport models are developed from, or have been shown to be consistent with, the same transport data. The primary differences between the transport models falls in three areas: whether the surface or substrate grain size is used, the number of size fractions used, and whether the model uses calibration (table 4.1).

**Table 4.1.** Differences between transport models.

| Model | Number of grain sizes | Surface or substrate | Calibration |
|---|---|---|---|
| PK (1982) | One | Substrate | No |
| PKM (1982) | Many | Substrate | No |
| Parker (90) | Many | Surface | No |
| WC (2003) | Many | Surface | No |
| Bakke (1999) | Many | Substrate | Yes |
| Wilcock (2001) | Two | Surface | Yes |

Whether using one, two, or many size fractions, or surface or substrate grain size, all of the models include a transport function giving transport rate as a function of excess stress $\tau/\tau_r$ (fig. 4.1) and some basis for estimating the reference stress $\tau_r$. In the Wilcock (2001) and Bakke and others. (1999) methods, the reference stress is estimated from transport observations. For the Parker-Klingeman (1982), Parker (1990a,b), and Wilcock/Crowe (2003) models, the reference stress for each grain size is given by a hiding function (fig. 2.7). Because each of the models represents the flow using shear stress $\tau$, part of the transport calculation involves determining $\tau$ from the discharge and channel input specified by the user. As discussed in Chapter 3, uncertainty in specifying $\tau$ is an important source of error in calculating transport rates. The approach used to estimate $\tau$ is presented after the discussion of each model.

# Models Incorporated in the Prediction Software

The complete formulas for each transport model are presented in the accompanying user's manual (Pitlick and others 2009). The discussion here outlines the models to highlight their dominant features and the salient differences among them.

## Substrate-Based Equations

The Parker-Klingeman-McLean equation (Parker and others 1982) is based on a single grain size, $D_{50}$, of the substrate. $W^*$ is calculated as a function of $\tau_{50}^*/\tau_r^*$ where $\tau_{50}^*$ is formed using $D_{50}$ of the substrate and $\tau_r^*$ is a reference Shields stress with a value of 0.0876. The transport equation of Parker and Klingeman (1982) predicts the transport of multiple size fractions. Dimensionless fractional transport rate $W_i^*$ is calculated as a function of:

$$\frac{\tau_{50}^*}{\tau_r^*}\left(\frac{D_{50}}{D_i}\right)^{0.018} \tag{4.1}$$

where, again, $\tau_{50}^*$ is formed using $D_{50}$ of the substrate and $\tau_r^*$ is a reference Shields stress with a value of 0.0876. The transport functions are plotted in figure 4.1 and the hiding function implied in Eq. 4.1 is shown in figure 2.7. Because these formulas are referenced to the substrate grain size, they are best applied to small gravel-bed streams with a modest sediment supply in a humid environment, similar to Oak Creek, Oregon, for which the formulas were originally derived.

## Surface-Based Equations

The surface-based transport model of Parker (1990a,b) was developed using the surface grain-size distribution and transport rates for Oak Creek. Because the surface grain-size distribution was available only for low flows, the relation was developed by inverting the PKM transport function, which is based on the bed substrate, and evaluating the result relative to the surface size distribution. A variation in surface grain size as a function of stage was inferred in the derivation of the relation. The Parker (1990a,b) transport model excludes grains smaller than 2 mm. Dimensionless fractional transport rate $W_i^*$ is calculated as a function of:

$$\omega \frac{\tau_{sg}^*}{\tau_{rsg}^*}\left(\frac{D_{sg}}{D_i}\right)^{0.0951} \tag{4.2}$$

where $\tau_{sg}^*$ is formed using $D_{sg}$, the mean size of the gravel portion of the bed surface, the reference Shields stress $\tau_{rsg}^*$ is given by Parker (1990a) as 0.0386, and $\omega$ is a straining function that depends on $\tau_{sg}^*/\tau_{rsg}^*$. Relations for $\omega$ are plotted in Parker (1990a) and can be found in tabulated form in Parker (1990b). The hiding function implied in Eq. 4.2 is nearly identical to that in the Parker-Klingeman relation (fig. 2.7).

The surface-based equation of Wilcock and Crowe (2003) was developed from coupled observations of flow, transport, and bed surface grain size using five sediments with varying sand content in a laboratory flume (Wilcock and others 2001). Unlike the Parker (1990a,b) surface-based model, sand is treated explicitly in the model. Dimensionless fractional transport rate, $W_i^*$, is calculated as a function of:

$$\frac{\tau}{\tau_{rsg}}\left(\frac{D_{sg}}{D_i}\right)^{\chi} \tag{4.3}$$

where the exponent $\chi$ varies with relative grain size $D_i / D_{sg}$ (fig. 2.7) and $\tau_{rsg}$ is the reference stress for the mean size of the bed surface. In the Wilcock-Crowe model, $\tau_{rsg}$ is not a constant but depends on $F_s$, the proportion of sand on the bed surface.

## Calibrated Transport Functions

The procedure of Bakke and others (1999) uses transport observations to calibrate the value of $\tau_r^*$ and the exponent 0.018 in Eq. 4.1. In BAGS, the calibration is conducted with least square regression subject to the constraint that the average transport rate calculated with the fitted equation is equal to the average transport rate from the samples. The procedure included in BAGS uses substrate grain size, although Bakke and others (1999) indicate that the same approach can be used with surface grain size.

The two-fraction equation of Wilcock (2001) is intended to be used in a calibration procedure in which values of the reference shear stress are chosen to match the formula with transport rate observations. Transport is grouped in two fractions, gravel and sand, the boundary of which is normally defined as 2 mm but can be defined at larger values in coarser beds (Wilcock and others 1997). Wilcock (2001) suggested that the choice of transport formula is of secondary importance when using transport observations to calibrate the formula. They suggested a formula for consistency with previous formulas, including the PKM formula. The calibration parameters in the model are the reference shear stresses for gravel and sand, $\tau_{rG}$ and $\tau_{rS}$, respectively. These are determined in BAGS using a least square regression on the gravel and sand transport data.

## Calculating Transport as a Function of Discharge

The transport formulas give transport rate as a function of shear stress, $\tau$. In order to calculate transport rates as a function of discharge, the software must determine $\tau$ as a function of user-specified values of discharge, channel geometry, and hydraulic roughness. This requires using a flow resistance equation (which relates $\tau$ to flow depth $h$, velocity $U$, and hydraulic roughness) and water mass conservation (which relates $h$ and $U$ to water discharge $Q$).

BAGS offers two different approaches for connecting $\tau$ and $Q$. In the first, the user specifies a value of Manning's roughness, $n$, for in-channel flows. Manning's equation and water mass conservation are then solved to find $U$, $h$, and the total boundary stress $\tau_0$. BAGS then performs a drag partition to estimate the grain stress $\tau'$, the portion of $\tau_0$ acting on the sediment grains. Derivation of the drag partition formula follows from the development in Chapter 2. Recall that the Manning formula can be combined with the depth-slope product to give:

$$\tau_0 = \rho g S_{1/4} (nU)_{3/2} \tag{2.13}$$

where $n$ is the value specified by the user. Recall, also, that we can estimate the portion of the total roughness $n$ due only to the bed grains using the Manning-Strickler formula (Eq. 2.11). We substituted this grain roughness $n_D$ in Eq. 2.13 to estimate the grain stress $\tau'$ in Chapter 2:

$$\tau' = \rho g S^{1/4} (n_D U)^{3/2} \tag{4.4}$$

One can think of the relation between Eqs. 4.4 and 2.13 as follows. If the only roughness in a channel were due to the bed grains, then $n = n_D$ and $\tau_0 = \tau'$. As other factors such as banks, in-channel obstructions, and bends add to the roughness, $n$ will exceed $n_D$ and $\tau_0$ will be larger than $\tau'$ for the same $U$ and $S$. Taking the ratio of Eqs. 4.4 and 2.13 gives:

$$\frac{\tau'}{\tau_0} = \left(\frac{n_D}{n}\right)^{3/2} \tag{4.5}$$

which is our formula for estimating the grain stress in BAGS. Although the relation between $n$, $n_D$, $\tau_0$, and $\tau'$ appears reasonable as we explain it, we cannot precisely show that the relation is exact for the same values of $U$ and $S$, as required to combine Eqs. 4.4 and 2.13 in Eq. 4.5. This problem has long been recognized in sediment transport modeling. Meyer-Peter and Müller (1948) and Einstein (1950) developed similar drag partition approaches in their classic models of sediment transport.

To use Eq. 4.5, a user-specified value of $n$ is needed. If such a value is not available, BAGS calculates the roughness from the bed material only and no drag partition is made. In general, this means that the transport rates estimated by BAGS are likely to be too large, with the amount of overestimate increasing for rougher channels. This is because an increase in hydraulic roughness (for a given $U$ and $S$) causes the depth to increase and the velocity to decrease and, thus, reduces the grain stress (Eq. 4.4). In general, transport calculations made without a drag partition will be most accurate if reach is wide, shallow, and mostly void of roughness elements other than the movable bed material. If a value of total roughness $n$ is not available but the Manning-Strickler $n_D$ for your bed material (calculated from Eq, 2.11) produces roughness values that seem much smaller than a reasonable value for the channel, some auxiliary BAGS calculations using trial values of $n$ can give an indication of the possible importance of including a drag partition in the transport estimate. It is worth noting again that the calibrated transport methods in BAGS provide, in essence, a drag partition, because the value of $\tau_r$ results from the calibration and the transport rates are calculated as a function of $\tau/\tau_r$. Thus, the problem of overestimating transport rates is most acute for the uncalibrated formulas used in BAGS.

Whether using a drag partition or not, BAGS must estimate the grain roughness from the bed material grain-size distribution. Although somewhat different relations have been proposed, all of the transport models correlate the roughness grain size with a coarser portion of the bed material size distribution. The roughness size for the Parker (1990a) relation is defined as $2D_{s90}$, where $D_{s90}$ is defined for the surface size distribution. Wilcock (2001) suggested estimating the roughness size as $2D_{s65}$. For the PKM and PK substrate-based relations, roughness height must be defined relative to the substrate size distribution. Based on limited field data in Parker and others (1982), a ratio of surface $D_{90}$ to substrate $D_{50}$ ranges between 4.3 and 6.4 with an average value of approximately 5.35. The roughness size recommended by Parker (1990a,b) is thus implemented for the substrate-based models as $10.7D_{sub50}$. Although there is variation in the different estimates for roughness grain size, the effect on the calculated hydraulics is generally small (recall from Eq. 2.11 than the grain roughness $n_D$ is proportional to grain size to the 1/6 power).

For a simple channel with a specified width of bed material, total transport rates can be calculated directly from the transport relations. In a compound cross section, BAGS calculates flow separately in the main channel and over the floodplain regions. Transport is calculated only in the main channel portion—the location of the bed material. Further detail on the hydraulic computations in BAGS is given in the user's manual (Pitlick and others 2009).

## Why a Menu of Models Can be Misused

Given a drop-down menu providing a choice of different transport formulas, it is tempting to select all the formulas in order to get some idea of the uncertainty in the calculated transport rate. This will, indeed, give a range of estimated transport rates, although it is hard to know what to make of it. The main source of uncertainty in calculated transport rates arises from uncertainty in the input values of grain size, boundary stress, and hydraulic roughness. Considerable effort has been spent over the years in comparing the accuracy of different transport formulas. Such comparisons inevitably suffer from the fact that the true transport rate is rarely known, but also divert attention from the primary source of error in calculated transport rates: uncertainty in the boundary conditions. Too often, the transport formula is blamed for poor results when the real culprit is poor input. A basis for estimating the uncertainty in calculated transport rate from uncertainty in the input is provided in Chapter 7.

# Chapter 5—Field Data Requirements

Although it is beyond the scope of this manual to give a detailed discussion of the field data needed to make transport estimates, we do present several factors to consider for collecting information to support development of a good estimate. An excellent description of basic field techniques is provided by Harrelson and others (1994). A comprehensive review of streambed sampling techniques is provided by Bunte and Abt (2001).

## Site Selection and Delineation

An essential factor in making an accurate transport estimate—probably the most important—is choosing an appropriate reach in which to work. Just as the USGS has developed guidelines for selecting reaches for stream gauging, there are attributes of a channel reach that will make it easier and more likely for you to develop a good sediment rating curve. The factors that contribute to error in transport estimates, summarized in Chapter 3, can be used to define the attributes.

### *The Flow Problem*

When the flow velocity and depth vary through a reach (for example, due to channel bends or changes in slope, flow constrictions and expansions), the non-uniform flow terms in the governing relation for open channel flow cause the total boundary stress $\tau_0$ to differ from that given by the depth-slope product (Eq. 2.7). BAGS calculates flow and transport at a cross-section and cannot account for non-uniform flow. If the non-uniform flow terms in Eq. 2.7 are significant, the calculated $\tau_0$ will be in error. A 1d hydraulic model (such as HEC-RAS) can account for the effect of non-uniform flow on reach hydraulics if the model is calibrated with observed stages at known discharges. Nonetheless, when $\tau_0$ varies along the channel, it is difficult to determine the spatially averaged value of $\tau_0$ that is best correlated with the total transport through the reach. Channel features that produce flow obstructions (vegetation, bridge abutments, boulders, and so forth) cause the grain stress $\tau'$ driving the transport to differ from the total stress $\tau_0$ in the reach. Although an expression for grain stress (Eqs. 2.14, 2.15) provides a basis for estimating $\tau'$, this method is an approximation and the accuracy of estimating $\tau'$ is likely to improve when $\tau'$ is a large fraction of the total stress.

Thus, an ideal reach for estimating transport rates is straight, with minimal variation in width, slope, bed material, or roughness and with nothing in the flow (other than the sediment) to take up the stress acting on the bed. This, of course, is an ideal that will rarely be exactly available. The physical constraints provide guidelines as to what to avoid in locating a sampling reach:

- Reaches with bends, changes in channel width, slope, or roughness that produce rapid changes in flow velocity and depth;

- Reaches with obstructions, such as boulders, vegetation, and debris jams, that exert substantial form drag causing the grain stress to be a reduced proportion of the total boundary stress; and

- Reaches with highly variable topography or bed material, such that the spatially averaged stress has an uncertain relation with the total transport through the reach.

Note that the attributes of a desirable reach for estimating transport with a formula are not necessarily the same as those for a reach well suited to the development of an empirical sediment rating curve. For example, a section in which the flow is forced through a constriction, such as a bridge or flume, can allow for very accurate flow and transport measurements, particularly if the bed is bedrock or concrete, allowing for secure placement of samplers on the surface (for example, Kuhnle 1992). Such locations are not appropriate for prediction using a formula—with or without calibration—because these formulas are developed from conditions in which the transport field is in a steady state with the flow and the bed from which the sediment is entrained.

### The Sediment Problem

Formula predictions are based on the grain size of sediment in the reach. If a reach is fully alluvial and at equilibrium such that the channel is formed entirely of material the stream is regularly transporting and the transport rates in and out of the reach are balanced over periods of a storm or longer, the sediment transport rates should be predictable as a function of the grain size in a reach. If the bed sediment is not in a steady state with the transport (for example, some coarser grains rarely, if ever are entrained), the predictions will be in error. Note that under steady state alluvial conditions, the grain size of the bed and transport will not generally be the same, because the transport formulas used in BAGS predict size selective transport. The important factors are that the transport regime be in a steady state with the current hydraulics and bed configuration and that all of the bed sediment be actively transported.

Of particular concern are cases in which coarser parts of the bed are effectively immobile under the present flow regime. This may happen below a dam or other site of sediment storage or where local sediment sources (such as hillslopes) introduce sediment to the channel that cannot be transported. If these grains are included in the bed material grain size, the rate of sediment transport is likely to be significantly underestimated.

Because adjustments in the stream bed can occur slowly, particularly in headwater streams with small transport rates, the best evidence of an adjusted alluvial reach is persistence in bed morphology and grain size over a period of years. Particular attention should be given to gravel bars. These are the locations where transported sediment is stored and are, therefore, more sensitive to changes in the sediment balance of a reach. If the channel experiences a range of floods that transport all of the bed material (indicating that there is not an inactive lag of immobile coarse grains) without substantial adjustments bar size or grain size (indicating that sediment inputs and outputs are balanced), application of a transport formula is appropriate.

### The Incipient Motion Problem

A range of factors can cause the critical Shields Number for incipient motion, $\tau_c^*$, to vary. Periods of intermediate flow producing small transport rates of coarse grains can produce a variety of bed structures at a range of scales (imbrication, clast jams, stone lines and cells). These structures can increase $\tau_c^*$ by a factor of two or more (Church and others 1998). Extended periods with little or no transport can promote fines infiltration, grain settlement and interlocking, biofilm growth, and chemical precipitation, all of which can increase the stress needed to initiate transport. Formula predictions using standard values of $\tau_c^*$ can be expected to work only with bed material that is entrained and transported on a regular basis. Nudging the bed with your toe should give an idea of whether the sediment is "loose," meaning that the resistance to movement is due to grain mass and relative size and not increased due to other factors.

## Channel Geometry and Slope

A channel cross-section and slope are needed to estimate the boundary stress from the water discharge. Particular care must be used when estimating the channel slope, because the mean change in elevation along a channel can be relatively small and often less than the change in elevation associated with the bed configuration and structure of the reach. Channel slope should be calculated from the change in elevation between comparable channel features such as riffles or bar tops. The width of the channel bed containing actively transported bed material should be noted. Guidelines and instructions for measuring channel geometry can be found in Harrelson and others 1994. If a hydraulic model, such as HEC-RAS has been calibrated for the reach, the water-surface slope from the model will give a more consistent slope estimate than the bed slope.

# Hydraulic Roughness and Discharge

An estimate of hydraulic roughness is needed to link water discharge, $Q$, to flow depth, slope, and boundary shear stress. The flow resistance and drag partition method presented here is based on the Manning equation, which requires an estimate of the roughness parameter $n$. This is not a trivial thing and uncertainty in $n$ introduces uncertainty in all subsequent calculations. The *only* reliable way to determine $n$ is to observe the water surface elevation (WSEL) at a known discharge. With a field survey of the cross-section geometry and channel slope, you can calculate the hydraulic radius, $R$, and the flow cross-sectional area, $A$, that are associated with the WSEL, as well as calculate $S$. These are used in Eq. 2.9 to back-calculate $n$. Unless your reach is in the immediate vicinity of a gage, the requirement of a known discharge means you will have to measure it. If you do not have a known discharge to match your WSEL observation, your estimate of $n$ is basically an educated guess. There are tables and picture books available to locate streams similar to yours to borrow $n$ values (Barnes 1967; Hicks and Mason 1998). There are formulas from which $n$ can be calculated as a function of sediment size, $R$, and sometimes other variables (Arcement and Schneider 1989). None of these methods are particularly accurate relative to directly measuring the discharge at a known stage. Guidelines and instructions for measuring discharge can be found in Harrelson and others 1994.

# Bed Material

Use of the Parker (1990a,b) and Wilcock/Crowe (2003) models requires specification of the surface grain size. Use of the Parker, Klingeman, and McLean (1982) and Parker-Klingeman (1982) formulas requires specification of the substrate grain size. Extensive guidelines for determining surface and substrate grain size are given by Bunte and Abt (2001). The surface grain size can be more variable than the subsurface grain size (Mueller and Pitlick 2005) and may require a larger number of samples. For these conditions, particular attention must be paid to developing a representative sample. Regions of the bed with similar bed material and hydraulics should be delineated and sampled separately. The reach averaged grain size is then developed using a weighted average of the grain size of different regions. Examples of facies mapping and reach-averaging grain size are found in Buffington and Montgomery (1999a,b).

A two-fraction approach can facilitate development of a reach-averaged estimate of grain size. If a characteristic grain size can be established for the fine and coarse portions of the streambed, areas with similar fines content can be mapped and combined to give a weighted average proportion of sand for the reach, giving an integral measure of grain size with reasonable effort. This provides a superior description of the bed compared to an unsupported extrapolation from detailed sampling at only a few locations.

## Sediment Transport

Samples of bed-material transport rate in gravel-bed rivers are most commonly collected using hand-held samplers such as the Helley-Smith (Emmett 1980). Although a number of criticisms of hand-held samplers have been raised (scooping non-moving bed material, perching on large grains, limited range of grain sizes), the most important limitation is simply that, short of an army of samplers simultaneously deployed for long periods of time, the samples collected by a hand-held sampler are very small relative to the large spatial and temporal variability associated with the bed-material transport. It is difficult (and unusual) for a hand-held sampler transect to collect as much as 1 percent of the transport passing the section over the sampling period. This amount is generally too small to provide a reliable estimate of bed material transport in gravel-bed streams. Even in the best of cases, scatter in transport observations is likely to exceed an order of magnitude and uncertainty in the predicted transport rate is difficult to reduce below ±50 percent (c.f. Hubbell 1987; McLean and Tassone 1987).

Superior sampling alternatives capable of collecting larger, more representative samples are in-bed pits (ranging from simple 5-gal buckets to sophisticated chambers with a load cell capable of tracking accumulated transport over time (for example, Church and others 1991; Reid and others 1980) or a net frame sampler fixed to the bed (Bunte and others 2004). At small transport rates, pit or net traps can sample for long periods such that the proportion of transport sampled scales directly with the proportion of the channel width covered by installed samplers. Pit or net samplers also offer the important advantage that, once installed, samples can be collected without the disturbance associated with placing and removing a hand-held sampler. These samplers have a number of drawbacks. An important problem is that these traps do not efficiently sample the transport of fine bed material. Sand and even fine pea gravel may not be efficiently trapped in a pit and the recommended mesh size for net samplers is 4 mm. Because fine bed-material transport tends to be more active and somewhat better mixed than coarse bed-material transport, a hand-held sampler may provide adequate sampling of the fine component. This suggests a hybrid approach in which each transport component—wash load, fine bed material load, and coarse bed material load—is sampled with a method best suited to its properties. In streams with significant organic load, the nets may quickly become clogged with organic debris.

USDA Forest Service RMRS-GTR-226. 2009.

The primary drawback of pit or net samplers is that they fill quickly at large transport rates, limiting their application to flows producing relatively small transport rates. As will be discussed in Chapter 6, if transport samples are to be used to calibrate a transport formula, the small-sample limitation is not a serious drawback.

# Chapter 6—Application

Making a useful transport estimate involves more than choosing a transport formula and plugging in the necessary input. One is typically interested in determining transport rate as a function of discharge, so a method for estimating the bed shear stress—the flow variable used in the transport formulas—is needed. This requires some description of the geometry, slope, and hydraulic roughness of the stream channel. The choice of grain size used in the transport calculation also involves some careful thought. This is not just a matter of choosing between a single characteristic grain size, two grain sizes, or many grain sizes. One must also choose between surface and substrate distributions. Further, one must develop an appropriate spatially averaged grain size for a reach and decide whether the reach is fully alluvial (such that the grain size can be determined from the bed material) or non-alluvial (such that the appropriate grain size is that of the load itself). This manual cannot provide all the answers for all situations. Our goal is to define the important problems and provide some perspective on how to address these problems.

The approach used when developing a transport estimate is also strongly constrained by the data available. We organize this chapter accordingly. The chapter begins with an overview of the general approaches possible—empirical versus formula versus calibrated. We then briefly summarize how an empirical relation can be developed before discussing how to use transport formulas with and without transport observations for calibration.

## Options for Developing a Transport Estimate

The typical choices for estimating bed-material transport in a gravel-bed river are to use a formula or directly measure the transport rate, typically with portable samplers. Broadly speaking, formula predictions require less effort, whereas field measurements offer the possibility of greater accuracy, but at greater effort. A third option, in which a few observations of transport rates are used to improve the accuracy of a formula prediction, may provide a superior combination of accuracy and effort.

An important advantage unique to the formula prediction is the ability to predict transport under conditions other than the present. If, for example, the rate and grain size of the sediment supply or channel geometry change in the future, the appropriate input to a formula prediction can be adjusted for the new conditions. In addition to discharge, the information required includes the grain size of the river bed and the channel geometry of the reach. Because the effectiveness of discharge in moving sediment varies with channel geometry, planform, and roughness, a suitably scaled flow variable (bed shear stress $\tau$ or stream power) is required for a formula prediction. Formulas based on $\tau$ are used in BAGS.

An empirical relation between transport rate and flow discharge may be developed directly from a large number of observations of the transport rate. Hand-held samplers can collect only a small proportion of the transport moving through the reach. Because the transport field can be highly variable in space and time, considerable scatter may be evident. The quality of the observations can be evaluated in terms of the coherence of the transport data collected at different times and flows. Net or pit samplers can provide good estimates of smaller transport rates but fill at larger transport rates. An empirical relation has the advantage of directly providing the desired relation between water and sediment discharge. But an empirical relation cannot be used to predict transport under future or altered conditions. Sampling campaigns are time consuming, expensive, and involve some risk if higher transport rates are to be measured.

A compromise between using a transport formula and developing an empirical sediment rating curve is to calibrate a transport formula using a small number of transport samples. If the number of samples used is small, thereby reducing effort, and the samples are accurate, thereby justifying the use of a small number of samples, a calibrated approach may provide substantially improved accuracy without the effort of a full sampling campaign. The calibration methods used in BAGS are used to determine the reference shear stress, $\tau_r$, a surrogate for the critical shear stress at incipient motion, $\tau_c$, used in the Parker-Klingeman (1982) and Wilcock (2001) transport models. Uncertainty in $\tau_r$ is a dominant source of error in transport predictions. As discussed in Chapter 3, calibration provides the necessary estimate $\tau_r$ and, in effect, performs the drag partition that connects the discharge to the appropriate level of grain stress producing the reference transport rate.

A brief consideration of the tradeoff between accuracy and effort suggests the possible advantages of a calibrated approach. A reasonable evaluation begins with two assumptions. First, the accuracy of the calibrated and formula approaches are the same when no transport samples are available because a formula is used to estimate the transport in both cases. Second, the accuracy of both the empirical and calibrated approaches increases with the square root of sample size, $N$ (consistent with the definition of confidence intervals for a regression line). With these two assumptions, Wilcock (2001) demonstrated that both the accuracy and ratio of accuracy to cost (a measure of value) for the calibrated approach should exceed those of either the formula or empirical methods and that a maximum value of the accuracy/cost ratio, should it exist, will occur at a smaller $N$ for the calibrated approach. As long as one assumes that the formula approach has some accuracy and that calibration of the formula prediction improves its accuracy, it appears reasonable to conclude that a few transport samples will provide increased accuracy at moderate cost and a favorable accuracy-cost ratio.

## How Many Grain Sizes?

Gravel-bed rivers are almost invariably characterized by a very wide range of grain sizes. Work over the past two decades has produced transport models that can predict the transport of many finely divided size fractions. Transport models based on the grain size of the bed surface are now available and have been included in BAGS (Parker 1990a,b; Wilcock and Crowe 2003). These formulas allow prediction of transport rates under transient conditions and realistic modeling of bed armoring. In practice, a transport model defined in terms of many size fractions requires a large amount of detailed input, to which the calculated result is sensitive. If these boundary conditions are poorly known, the transport prediction will be poor, no matter how good the model. Many-fraction models are appropriate for scenario evaluation and design testing and essential if changes in grain size are to be predicted (for example, in response to changes in sediment supply or bed armoring). But, because they are fragile—sensitive to uncertainties in boundary conditions—they are difficult to apply to the basic task of predicting transport rates at a particular location at a particular time. For this task, the tool needed should be robust, giving reliable results even with uncertain input, because in practice the information available for prediction is rarely sufficient.

If one argues that a model using fewer size fractions will be more robust, the most robust transport model would use a single size fraction. A single fraction model based on the Parker and others (1982) analysis of the Oak Creek transport data (Milhous 1973) is provided in BAGS. Although transport predictions using a single characteristic grain size have long been used, this approach allows for no change in the grain size of the bed or transport. It does not account for the common observation that finer fractions (typically sand) are actively transported at flows unable to move the larger gravel fractions that make up the framework of the stream bed (Carling 1989; Emmett 1976; Jackson and Beschta 1982). This distinction between the transport of fine and coarse bed material load suggests that a model based on two size fractions—sand and gravel—might offer both conceptual and practical advantages. A two-fraction estimate allows sand and gravel to move at different rates, thereby permitting change in bed grain size due to changes in the relative proportion of sand and gravel, but not due to the changes in the representative grain size of either fraction. This provides a means of predicting the variation in the fines content of the bed, which may often be more variable than that of the coarse fraction, and whose passage, intrusion, or removal may be a specific environmental or engineering objective. A two-fraction approach also provides a ready means of representing the interaction between the fine and coarse components of the bed material. The two fraction transport model of Wilcock (2001) is included in BAGS.

## Empirical Sediment Rating Curves

Most sediment transport problems require an estimate of the sediment transport rate $Q_s$ as a function of water discharge $Q$. A relation giving $Q_s$ as a function of $Q$ is called a **sediment rating curve**. A sediment rating curve typically takes the form of a power function:

$$Q_s = aQ^b \tag{6.1}$$

where, in the United States, $Q_s$ is in units of tons per day and $Q$ is in units of ft$^3$/s, or cfs. Preferable units would be kg/hr or metric tons per day and m$^3$/s.

As an illustration, consider the calculation of mean annual sediment yield. Using a sediment rating curve such as Eq. 6.1 and a record of discharge (for example, the daily mean $Q$ for 25 years), the total sediment load (the sediment yield) is determined by using Eq. 6.1 to calculate the tons of sediment transported for each day and then adding up all 9,131 or so values to get a total sediment yield for 25 years. Dividing this value by 25 then gives the mean annual sediment yield. Two important issues should be kept in mind when making such a calculation. First, you can't assume that a sediment rating curve will remain the same over time. Changes in sediment supply, channel configuration, or channel bed material can change the coefficients and perhaps even the form of the rating curve. Second, discharge may change rapidly on many streams (for example, small snowmelt-dominated streams; flashy arid and urban streams), such that daily mean discharge will not accurately represent the flow producing transport. In a small watershed, an afternoon thunderstorm might produce a flood that comes and goes in a few hours. In such cases, you will need a discharge record with a much finer time resolution (for example, 15 minutes). It is worth noting that this "flashiness" problem also applies to the development of a sediment rating curve from formulas.

If we directly and simultaneously measure both $Q_s$ and $Q$, we can, with enough observations, develop an empirical sediment rating curve, which usually amounts to determining the slope $b$ and intercept $\log(a)$ when fitting a straight line to:

$$\log(Q_s) = \log(a) + b\log(Q) \tag{6.2}$$

Exponentiating Eq. 6.2 gives Eq. 6.1. This is easy to do in a spreadsheet. A common alternative sediment rating curve is:

$$Q_s = a(Q - Q_c)^b \tag{6.3}$$

where $Q_c$ is the discharge at which substantial transport begins. The slope $b$ and intercept $\log(a)$ are determined by fitting a straight line to:

$$\log(Q_s) = \log(a) + b\log(Q - Q_c) \tag{6.4}$$

and then exponentiating. In a spreadsheet, this is done by exploring different values of $Q_c$ until a best fit is achieved, which could be judged by a minimum in the sum of squared errors about the fitted line.

When fitting a sediment rating curve, the data used should be reviewed very carefully. One or two outliers can have a strong effect on the fitted relation, particularly if the total number of data points is small. Exclusion of these points may be justified by examining the residuals (difference between fitted and observed). If the residual is more than two or three standard deviations from the mean residual, consider excluding that point. In some cases, expert advice may be neededz. Neglecting outliers is a debated topic and there are other issues involved in fitting a regression line to log-transformed data. In all cases, values not used in fitting the rating curve must be reported with the rest of the data. Common sense can help avoid egregious error. One can ask, for example, if the fitted value of $Q_c$ is plausible (it should correspond to a discharge at which some, but not too much sediment is moving). There is no substitute for examining a plot of the data with the fitted sediment rating curve.

Direct measurements provide "the real thing"—a relation between $Q_s$ and $Q$. Although the error in an empirical sediment rating curve can be large, it is generally better constrained than the error possible when using a formula. The tradeoffs are that the measurements needed require considerable effort and expense, introduce safety concerns, and, by their empirical nature, provide no certain ability to predict transport under conditions other than those measured. Because of the necessary effort, empirical sediment rating curves can only be developed for a small number of sites. To estimate transport at many sites, or throughout a watershed, or to predict transport under future or altered conditions, one needs a predictive model. Barry and others (2004, 2005) have proposed a method for predicting sediment rating curves in which the coefficient and exponent are functions of channel and watershed characteristics.

## Formula Predictions

Prediction of transport rate for a specified water discharge requires an estimate of the stress $\tau$ acting on the stream bed, along with grain size and width of the river bed. As discussed in Chapter 2, the grain stress, $\tau'$ (the portion of $\tau'$ acting on the moveable grains), cannot be calculated precisely, but must be estimated using a drag partition formula. An estimate of grain stress $\tau'$ requires an input slope and hydraulic radius. To get hydraulic radius from discharge, one must specify cross-section geometry, slope, and roughness. Stage and hydraulic radius are found from flow resistance and mass conservation relations. For a non-calibrated transport formula, $D$ is found from the bed surface or substrate, depending on the model. Using a calibrated approach, $D$ is determined from the grain size of the transport. A summary of the necessary information is given in table 6.1.

**Table 6.1.** Information needs for different transport estimates.

| Information required | Substitute | Uncalibrated formula | Calibrated formula | Empirical |
|---|---|---|---|---|
| Channel cross-section | Top width | ● | ● | ○ |
| Energy slope, water slope | Bed slope | ● | ● | ○ |
| Discharge | | ● | ● | ● |
| Roughness | | ● | ● | ○ |
| Bed Material grain size | | ● | ± | |
| Transport rate & grain size[‡] | | - | some | many |

● Required data. ○ Data useful for auxiliary computations.

± Wilcock (2001) requires no bed material grain size data; Bakke and others (1999) requires a bed grain-size distribution.

[‡] Development of the model also requires a discharge measurement or estimate for each transport measurement.

## Which Formula?

One of the more confusing issues for those using software to calculate transport rates is deciding which formula to use. A common recommendation concerning choice of transport model is to use a formula that was developed for conditions similar to those that you will model. Although correct in principle, this advice is not often helpful because many transport models are developed from much the same data, or have been shown to fit the same data sets. This is the case for the transport models in BAGS: the models due to Parker are based largely on the transport data from Oak Creek, Oregon (Milhous 1973). The surface-based model of Wilock/Crowe is based on a flume data set and was subsequently shown to fit the Oak Creek data (Wilcock and Crowe 2003; Wilcock and DeTemple 2005). All of these data are for the transport of mixed-size sediment, ranging in size from sand to medium to coarse gravel, transported over a bed with negligible to modest topography.

Not surprisingly, the transport functions at the heart of each model are quite similar, which is evident if all of the transport functions are placed on the same plot (fig. 4.1). The fundamental differences between the transport models concerns the data used to make the calculation. The key issues are whether:

1. transport observations are available to calibrate the estimate;

2. surface or substrate grain size is used; and

3. one, two, or many size fractions are used.

If reliable transport observations are available, the Wilcock (2001) model can be used to model two fraction (sand and gravel) transport rates and the Bakke (1999) procedure can be used to model the transport of many size fractions. Calibration is likely to provide a superior estimate over using a formula alone for two reasons. First, using the grain size of the sediment in transport eliminates the need to determine whether the bed grain size in the reach is well correlated with the transport. Second, by fitting the transport function at one flow (the flow producing the reference transport rate), the calibration performs the drag partition (it determines the grain stress) at that flow. The model then needs to predict the variation with discharge of the grain stress relative to the calibrated value. It is far easier to accurately estimate the change in grain stress than its actual value at any particular flow.

Transport estimates using a formula without calibration require specification of the grain size in the channel bed. The transport entrained by the flow at any moment depends on the grains available for transport—the grains on the bed surface. Hence, a surface-based transport model provides a more representative estimate of transport rate. It is also desirable because the bed grain size often available is from a pebble count—a measure of the surface size distribution. Substrate-based transport models were initially developed because information on the bed surface during transport measurements was not available. Transport observations in flumes were scaled by the sediment placed in the bed or fed into the flume. Transport observations in the field used the bed substrate because observations of the bed surface during active transport were not (and are still not) available. The conceptual problem with a substrate transport model is that it must implicitly account for the surface sorting (on which the actual transport rate depends). Because a variety of factors—most importantly flow history and sediment supply—may influence the degree to which the surface grain size differs from the substrate, the same substrate size distribution could be associated with different surface size distributions and, therefore, different transport rates even though a substrate-based transport formula will predict the same transport rates in each case.

Although a surface-based transport relation is physically correct, an important question concerns its variation over a hydrograph. If the bed surface grain size can only be measured at low flows and the surface changes (for example, becomes finer grained) over a flood, then a constant surface grain size will not accurately predict the transport over the flood. Although we do not have surface grain size observations during substantial transport rates (the closest is due to Andrews and Erman [1986] during a snow-melt flood of low frequency but not exceptionally large $\tau^*$), two recent computational studies suggest that an armor layer observed at low flows may actually persist throughout a flood hydrograph (Parker and others 2006; Wilcock and DeTemple 2005). Although these results do not prove that armor layers persist throughout floods in all streams, they do indicate that a change in surface grain size is not necessarily driven by properties inherent in the transport processes.

It is known that the degree of armoring varies widely from stream to stream. Hence, application of a substrate-based transport model can be expected to be in error to the extent that armoring in your stream differs from that in Oak Creek, on which the PK and PKM substrate-based models are based. Substrate-based transport models should be applied only for gross, overall predictions (Parker 2008).

The choice of a transport model should be consistent with the data available. If surface grain size data are available, the P90 or WC relations are appropriate. If substantial sand (more than 5 to 10 percent on the bed surface) is present, the WC model, which explicitly accounts for the effect of sand on gravel transport rates, is advantageous. If substrate data are available, then the PKM or PK models should be used. If you must decide which data to collect, surface grain size data are easier to collect and the application of a surface-based model has a stronger foundation.

The final choice concerns the number of size fractions to use. The PKM model predicts the transport rate of one size—the median size of the substrate—and thus cannot provide any indication of size-selective transport rates. It is best applied to gravel beds containing a small amount of sand (<10 percent) in order to correspond to the conditions for which the model was developed and because it does not allow for an explicit calculation of the effect of sand content on the transport rate. The PK, P90, Bakke, and WC models predict transport rates for many size fractions. In most cases, specifying grain size in $1\psi$ increments is sufficient to capture most of the variation of transport rate with grain size. With the exception of the Bakke method (because it uses transport samples to calibrate the prediction), the accuracy of these models is sensitive to the accuracy of the surface size distribution. If the surface grain size is poorly known, or if there is considerable topography and spatial variability in surface grain size, it is difficult to determine the correct input size distribution for accurate estimates of the transport rate.

A two fraction model provides a useful alternative because it more readily supports a spatially integrative measure of surface grain size (Chapter 5). Use of a calibrated transport model is an advantage with respect to questions regarding the choice of grain size: surface *v.* substrate, one *v.* two *v.* many size fractions. A calibrated model uses the grain size of the transport, which is sampled, rather than the grain size of either surface or substrate and it can be divided into as many fractions as desired.

# Chapter 7—Working With Error in Transport Estimates

## Assessing Error in Estimated Transport Rates

When using formulas to estimate transport rates, uncertainty arises primarily from error in the input. If uncertainty in the input variables can be described, this error can be "propagated" through the transport calculations. In some cases, it is possible to evaluate this error analytically. We present here a simpler and flexible approach: a Monte Carlo error analysis of error propagation. A spreadsheet using Monte Carlo analysis to calculate uncertainty in transport estimates, "MonteCarloTransport.xls," is available from the National Center for Earth-surface Dynamics (http://www.nced.umn.edu/Stream_Restoration.html). The program calculates uncertainty in the critical discharge, $Q_c$, at which sediment transport begins and the cumulative bed material transport rate over a hydrograph.

The idea underlying a Monte Carlo error analysis is straightforward. First, you specify a distribution representing the range of possible values for each input variable with uncertainty. Then, you choose a set of input values drawn from those distributions and make your calculation. This step is repeated many times, each time drawing a new set of input values from their distribution. The number of calculations made is typically very large (for example, 1,000 or more), such that the distribution of calculated values is sufficiently stable. The final step is to use the distribution of calculated values to represent the variability that is possible due to uncertainty in the input.

The first task is specifying uncertainty in the input parameters. Suppose that your best guess of Manning's *n* for your section is 0.028 and that you are 95 percent certain that the true value lies somewhere between 0.024 and 0.032. Your goal is a distribution of *n* values that represents your estimate of the real value. Reasonably, the real value is more likely to be closer to 0.028 (your best guess) than 0.024 or 0.032. If you represent the distribution of *n* values with a normal distribution with mean 0.028 and standard deviation 0.002, about 95 percent of the area under the probability distribution falls between 0.024 and 0.032. This distribution is shown in figure 7.1.a.

In figure 7.1.b, we express the same distribution in a cumulative form—we are cumulating the area under the curve in figure 7.1.a from left to right. The cumulative distribution in figure 7.1.b illustrates how the Monte Carlo input works.

We randomly choose a number from a uniform distribution between 0 and 1 (such a function is routine in most numerical computer programs, including Microsoft Excel) and then use figure 7.1.b to find the value of *n* associated with that random number. Because the cumulative curve is steeper in the vicinity of 0.028 (corresponding to the center of the distribution in fig. 7.1.a), the number of *n* values close to 0.028 will be much greater than from the "tails" of the distribution.

This process is repeated for all other input variables with uncertainty. Once a set of input values is selected, the transport calculation is made. The same process—determining values for each input variable and calculate transport—is repeated many times (1,000 times in the spreadsheet model). So, if you can specify a distribution for each input variable with uncertainty, the result is a distribution of calculated transport rates that will reflect the likely distribution—*the uncertainty*—in the input.

**Figure 7.1.** Illustration of input selection in Monte Carlo estimate of error propagation.

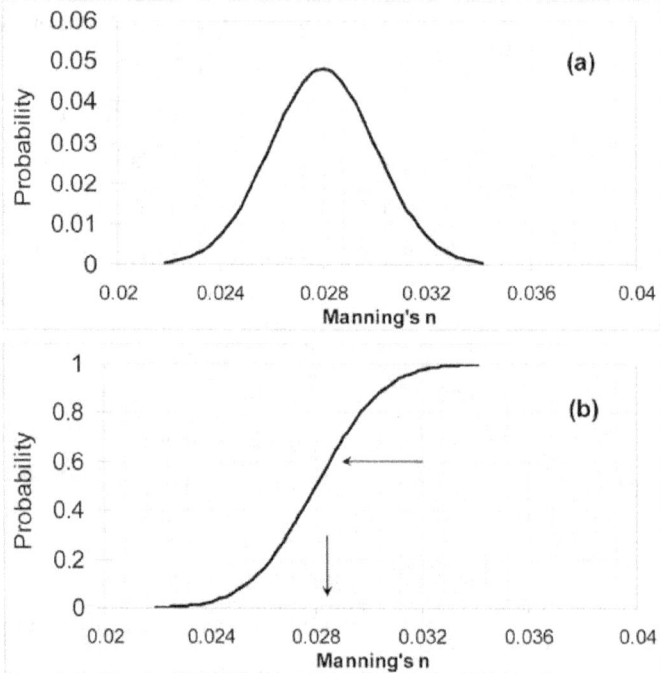

The two calculations in *MonteCarloTransport.xls* (critical discharge for incipient motion and cumulative transport over a hydrograph) use relations defined in terms of the grain stress, $\tau'$. To connect $\tau'$ to the discharge, the program uses simple hydraulic relations for water mass conservation, channel geometry, flow resistance, and shear stress in a wide, steady uniform flow:

$$Q = BhU \qquad (7.1)$$

$$B = \alpha Q^{\beta} \qquad (7.2)$$

USDA Forest Service RMRS-GTR-226. 2009.

$$U = \frac{\sqrt{S}}{n}h^{2/3} \tag{7.3}$$

$$\tau = \rho ghS \tag{7.4}$$

where $B$ is channel top width, $h$ is mean channel depth, $U$ is mean channel velocity, $\alpha$ and $\beta$ are parameters representing channel cross-section shape, $S$ is channel slope, $\rho$ is water density, and $g$ is the acceleration of gravity. Because our goal is to develop an idea of the uncertainty in the estimate, for these calculations, we choose to use a very simple definition of channel geometry. This makes the estimate easier to understand and interpret. You should be able to come up with a better estimate of $Q_c$ or transport rate, but as long as the range of input values used in the Monte Carlo analysis are similar (similar grain size, slope, channel width, discharge) to your problem, the range in the calculated values relative to the mean or median value should be similar.

To calculate uncertainty in the critical discharge for incipient motion $Q_c$, Eqs. 7.1 through 7.4 are solved for $\tau = \tau_c$ where $\tau_c$ is given using a specified value of the critical Shields Number $\tau_c^*$ in:

$$\tau_c^* = \frac{\tau_c}{(s-1)\rho gD} \tag{7.5}$$

In developing this estimate, the primary sources of uncertainty are taken to be the value of $\tau_c^*$ (which may vary as the result of a number of factors such as the development of armor and surface structures; see Chapter 3—The Incipient Motion Problem), the representative grain size $D$ (due to sampling uncertainty and the choice of $D$ from the bed material size distribution; see Chapter 3—The Sediment Problem), and the appropriate value of Manning's $n$ (see Chapter 2—The Drag Partition and Chapter 3—The Incipient Motion Problem). A Monte Carlo simulation incorporating uncertainty in $\tau_c^*$, $D$, and $n$ demonstrates the uncertainty in the calculated value of $Q_c$. An example is illustrated in figure 7.2, which uses typical values of the input parameters with modest estimates of their uncertainty. Parts a, b, and c of figure 7.2 show the 1,000 values of $\tau_c^*$, $D$, and $n$ used in the calculations. Part d shows the 1000 calculated values of $Q_c$. Approximately 90 percent of the calculated values of $Q_c$ fall within a factor of four: roughly between 4 and 16 m³/s. This discouraging result is made worse by the fact that the example does not include other important problems in determining $Q_c$, including the fact that variable channel topography and bed material grain size will cause some grain sizes in some locations to begin moving at flows for which the mean values of $\tau$ and $U$ (as used above) indicate that $Q < Q_c$.

The intent of this example is not to discourage calculation of $Q_c$ so much as it is to discourage belief in the calculated values. What can be done? Uncertainty in Manning's $n$ can be nearly eliminated with accurate measurements of discharge, although uncertainty in $\tau_c^*$ and $D$ is harder to reduce. Is there an alternative? If actual values of $Q_c$ are needed, the certain (and relatively simple) approach is to observe the displacement of tracer gravels, provided that it is possible to visit the site following a range of different discharges. Also, in some cases it may be possible to pose the problem in relative rather than absolute terms (for example, the change in $Q_c$ due to a change in grain size can be estimated more reliably than $Q_c$ itself). This is discussed further in the next section.

Uncertainty in calculated transport rates is calculated using the Meyer-Peter and Müller equation:

$$\frac{q_s}{\sqrt{(s-1)gD^3}} = 8\left(\frac{\tau}{(s-1)\rho gD} - \frac{\tau_c}{(s-1)\rho gD}\right)^{3/2} \tag{7.6}$$

Although a more recent transport formula could be used, M-PM offers the advantages of familiarity and simplicity. Because the overall form of most transport formulas are not strikingly different, the range of uncertainty may be expected to be similar regardless of the formula used in making the uncertainty estimate. As for the calculation of $Q_c$, the estimate of transport rate accommodates uncertainty in $n$, $\tau_c^*$, and $D$. The reasons for the uncertainty are the same, with the additional uncertainty that the grain size of the transported sediment may be underrepresented in the study reach. The transport rate at a given $Q$ is calculated using Eq. 7.6 with Eqs. 7.1 through 7.4 to solve for $\tau$. Transport rates are calculated as the total over a hydrograph. For the hydrograph shown in figure 7.2.e, the 95 percent prediction interval is 43 percent of the cumulative transport calculated using the mean values of $n$, $\tau_c^*$, and $D$, indicating considerable uncertainty in the calculated result.

It is argued that the effect of uncertainty in predicted transport rates is reduced in sediment balance calculations (for example, Soar and Thorne 2001) because the balance is defined as the difference in transport rates between adjacent reaches and it is assumed that some of the error in the calculations in each reach will cancel. A stochastic estimate of transport (as in fig. 7.2f) applied to each reach can provide an estimate of the uncertainty in the calculated sediment balance. The calculations shown in figure 7.2 are quite simple to make in a spreadsheet, such that it is not difficult, and should be common practice, to report the uncertainty and, where possible, incorporate it in channel design.

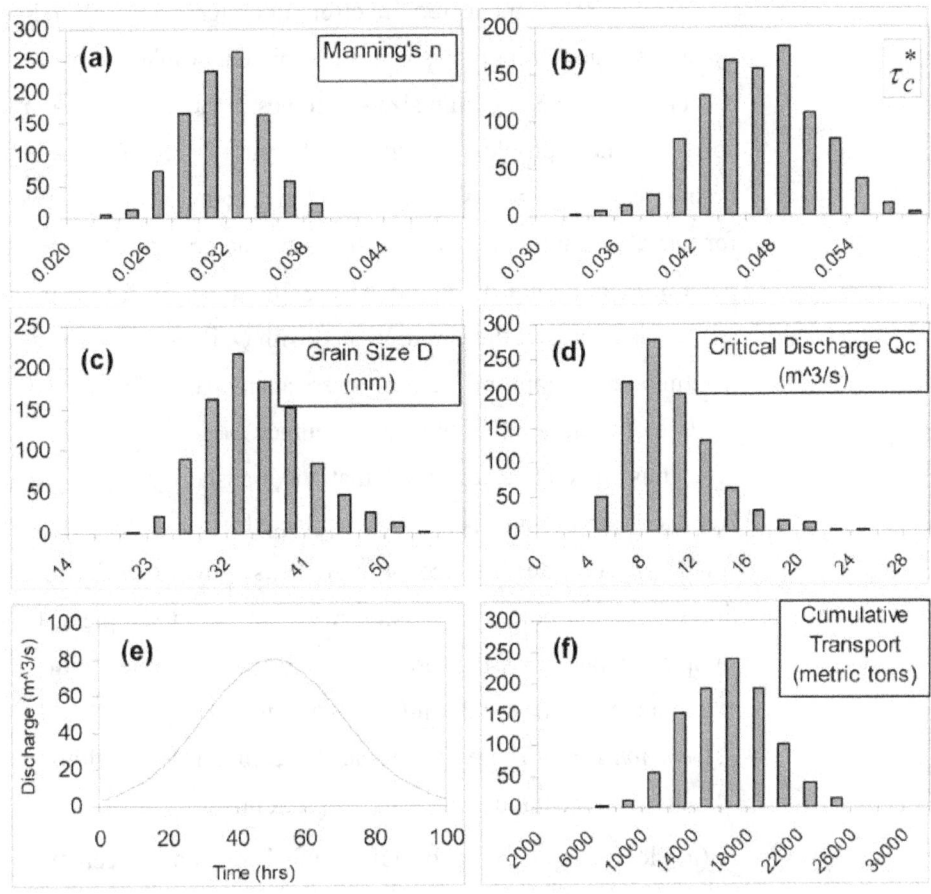

**Figure 7.2.** Monte Carlo estimate of uncertainty in calculated critical discharge $Q_C$ and transport rate based on 1,000 trials drawn from specified uncertainty in (a) Manning's $n$, (b) critical Shields Number, and (c) characteristic grain size $D$. (d) Calculated $Q_C$. (f) Calculated cumulative transport, using the hydrograph in (e) and active transport width of 5 m. Calculations use channel slope $S$ = 0.003, channel width $B$ (m) = $5Q^{0.1}$. Grain size $D$ (mm) = $2^\psi$.

## Strategies

Although there is no simple solution to dealing with uncertainty in transport calculations, some strategies for working with that uncertainty can be offered.

1. *Calibration.* Using a few transport samples to calibrate your transport estimate is the single most effective thing you can do to increase accuracy. The same problem applies to the prediction of $Q_c$ and transport rate: under typical conditions, uncertainty in boundary conditions is sufficiently large that the calculated results of a formula have very large uncertainties. If a good estimate is required, it must be determined from field observations. This is no different from the requirements for accurately determining the hydraulic roughness. If a reliable value of Manning's $n$ is needed, discharge measurements are required.

Field measurements require effort, although the payoff in improved accuracy can be large. Calibration solves the sediment problem (Chapter 3) because you do not need to specify grain size—it comes in the transport samples. Calibration solves the flow problem (Chapter 3) because it provides an accurate value of grain stress $\tau'$ at one discharge (the $Q$ producing $\tau_r$). This value of $\tau'$ accounts for spatial averaging, flow non-uniformity, and drag partitioning, all of the messy parts of determining the appropriate grain stress for a transport problem. One still needs to predict the variation of $\tau'$ with $Q$, but this is a much easier task than determining the appropriate grain stress at any particular value of $Q$ (see Chapter 3—Use of Calibration to Increase Accuracy).

It is, of course, essential that the transport samples be quite accurate. Accurate samples are likely to be collected using pit or net frame samplers at small transport rates because these methods allow for collection of a larger fraction of the transport than possible with hand-held samplers. It is also safer to sample smaller flows that produce small transport rates with larger frequency, which increases the opportunity for sampling.

2. *Common sense.* There are no rules for common sense, although there are a number of techniques that can help it flourish.

   (a) Plot the input to the formulas—this is the most effective way to catch error.

   (b) Plot the prediction as transport rate versus discharge. Does it pass the laugh test? Is there sediment moving at implausibly small flows? Is there no sediment moving at floods?

   (c) Make field observations following higher flows. Is there evidence of fresh entrainment and deposition of bed material? Indications of recent transport include the number and location of clasts with clean surfaces relative to those with established biofilms or staining, loose streambed, fresh deposition on the top of vegetation or upstream of debris jams, and changes in bar geometry. If most of the bed appears to have been actively transported, was the magnitude of that flood consistent with the discharge at which predicted transport begins?

   (d) Order of magnitude evaluation. For example, is a calculated transport rate so large that it would fill a large reservoir in a year or two? Or so small that a local deposit observed behind a debris jam would have required a century to form?

3. *Extrapolation.* For an empirical or calibrated approach, the further the discharge from the range of discharges with observations, the more uncertain your estimate. Of particular importance is when flows go out of bank, in

which case the relation between discharge and $\tau$ can change. Extrapolation for in-channel flows should be more reliable than for out-of-channel flows.

4. *Reframe the question*. In many practical problems, the obvious thing to do is to calculate the transport rate. That's what BAGS is for, right? Yes and no. As we have discussed at length, an estimate of transport rate includes considerable error. In contrast, the difference between two calculated transport rates, or the ratio of two calculated transport rates, may have far less error. The reason is that accuracy in estimating differences or ratios depends only on being "in the ballpark" such that rates of change in transport rate are reasonably well captured. Because the underlying transport functions are nonlinear, you still have to be close, but the sensitivity of the answer to the accuracy in the input will be lower.

As an illustration, if one were designing a channel that was to have negligible transport at a design flow $Q_d$, the design criteria could be defined as having a small probability of $Q_d > Q_c$. Thus, one would perform calculations such as in figure 7.2d with the goal that $Q_d$ would fall toward the few smallest values of the calculated $Q_c$. Evaluation of different channel designs (different channel geometry, different grain size) would be based on the shift of the calculated $Q_c$ distribution relative to $Q_d$.

As another example, if one is concerned with adjustments in a stream reach due to changes in the watershed that affect sediment supply, one might reasonably calculate transport rates for both the "before" and "after" conditions. The confidence that can be placed in their difference, or ratio, will be larger than can be placed in either value individually. Similarly, if one wished to evaluate transport through a reconstructed stream reach, the confidence that can be placed in an estimate of transport rate in the new reach at any particular flow will be smaller than the confidence one could place on the difference in transport rate between the upstream (supply) reach and the reconstructed reach. Because it is the difference in transport rates, rather than their actual values, that determine whether the new reach will store or evacuate sediment, the relative and easier to answer question is actually the quantity of primary interest.

When asking questions about changes in transport relative to changes in the controlling variables (for example, how much would transport rate change if the grain size were half the size?), ratios of calculated transport rates are often the most useful. When asking questions that depend on actual amounts of transport (for example, whether a reach will fill or empty of sediment and how fast), difference of calculated transport rates are most relevant.

# References

Andrews, E. D.; Erman, D. C. 1986. Persistence in the size distribution of surficial bed material during an extreme snowmelt flood. Water Resources Research. 22: 191-197.

Arcement, G. J., Jr.; Schneider, V. R. 1989. Guide for selecting Manning's Roughness Coefficients for natural channels and flood plains. U.S. Geological Survey. Water-supply Paper 2339.

Bakke, P. D.; Baskedas, P. O.; Dawdy D. R.; Klingeman, P. C. 1999. Calibrated Parker-Klingeman model for gravel transport. Journal of Hydraulic Engineering. 125(6): 657-660.

Barnes, H. H., Jr. 1967 Roughness characteristics of natural channels. U.S. Geological Survey. Water-Supply Paper 1849.

Barry, J. J.; Buffington, J. M.; King, J. G. 2004. A general power equation for predicting bed-load transport rates in gravel bed rivers. Water Resources Research. 40(10): W10401. doi:10410.11029/12004WR003190.

Barry, J. J.; Buffington, J. M.; King, J. G. 2005. Reply to comment by C. Michel on: A general power equation for predicting bed load transport rates in gravel bed rivers. Water Resources Research. 41: W07016. doi:10.1029/2005WR004172.

Borland, W. M. 1960. Stream channel stability. United States Bureau of Reclamation. Denver.

Brownlie, W. R. 1981. Prediction of flow depth and sediment discharge in open channels. Report No. KH-R-43A. W. M. Keck Laboratory of Hydraulics and Water Resources. California Institute of Technology. Pasadena, California. 232 p.

Buffington, J. M.; Montgomery, D. R. 1997. A systematic analysis of eight decades of incipient motion studies, with specific reference to gravel-bedded rivers. Water Resources Research. 33(8): 1993-2029.

Buffington, J. M.; Montgomery, D. R. 1999a. A procedure for classifying textural facies in gravel-bed rivers. Water Resources Research. 35(6): 1903-1914.

Buffington, J. M.; Montgomery, D. R. 1999b. Effects of hydraulic roughness on surface textures of gravel-bed rivers. Water Resources Research. 35(11): 3507-3522.

Bunte, K.; Abt, S. T. 2001. Sampling surface and subsurface particle-size distributions in wadable gravel- and cobble-bed streams for analyses in sediment transport, hydraulics and streambed monitoring. Gen. Tech. Rep. RMRS-GTR-74. Fort Collins, CO: U.S. Department of Agriculture, Forest Service, Rocky Mountain Research Station. 428 p. www fs fed.us/rm/pubs/rmrs_gtr74. html.

Bunte, K.; Abt, S. R.; Potyondy, J. P. 2004. Measurement of coarse gravel and cobble transport using portable bedload traps. Journal of Hydraulic Engineering. 130(9): 879-893.

Carling, P. A. 1989. Bedload transport in two gravel-bedded streams. Earth Surface Processes and Landforms. 14: 27-39.

Church, M.; Hassan, M. A. 2002. Mobility of bed material in Harris Creek, Water Res. Res. 38(11): 1237. doi:10.1029/2001WR000753.

Church, M.; Hassan, M. A.; Wolcott, J. F. 1998. Stabilizing self-organized structures in gravel-bed stream channels. Water Resources Research. 34: 3169-3179.

Church, M.; Wolcott, J. F.; Fletcher, W. K. 1991. A test of equal mobility in fluvial sediment transport: behavior of the sand fraction. Water Resources Research. 27(11): 2941-2951.

Clark, J. J.; Wilcock, P. R. 2000. Effects of land use change on channel morphology in northeastern Puerto Rico. Bulletin, Geol. Society of America. 112(12): 1763-1777.

Curran, J. C.; Wilcock, P. R. 2005. The effect of sand supply on transport rates in a gravel-bed channel. J. Hydraulic Engineering. DOI: 10.1061/(ASCE)0733-9429(2005)131:11(961).

Einstein, H. A. 1950. The bedload function for sediment transport in open channel flows. Tech. Bull. 1026. U.S. Dept. of Agriculture, Soil Conserv. Serv. Washington, DC.

Emmett, W. W. 1976. Bedload transport in two large, gravel-bed rivers, Idaho and Washington. Proceedings of the third federal inter-agency sedimentation conference, Denver. Sediment. Comm. Water Resour. Council. Washington, DC: 4-101-114.

Emmett, W. W. 1980 A field calibration of the sediment-trapping characteristics of the Helley-Smith bedload sampler. USGS Professional Paper 1139.

Gilvear, D.; Bryant, R. 2003. Analysis of aerial photography and other remotely sensed data. In: Kondolf, G. M.; Piegay, H. (eds.). Tools in fluvial geomorphology. John Wiley & Sons.

Harrelson, C. C.; Rawlins, C. L.; Potyondy, J. P. 1994. Stream channel reference sites: an illustrated guide to field technique. Gen. Tech. Rep. RM-245. U.S. Department of Agricluture,

Forest Service, Rocky Mountain Forest and Range Experiment Station. File: RM245e.pdf (5,175KB/67p. available:http://www.stream.fs fed.us/publications/documentsStream.html)

Haschenburger, J. K.; Wilcock, P. R. 2003. Partial transport in a natural gravel-bed channel. Water Resources Research. 39(1): 1020. doi:10.1029/2002WR001532.

Hassan, M. A.; Church, M. 2000. Experiments on surface structure and partial sediment transport on a gravel bed. Water Resources Research. 36(7): 1885-1985.

Henderson, F. M. 1966. Open channel flow. McMillan. New York.

Hicks, D. M.; Mason, P. D. 1998. Roughness characteristics of New Zealand Rivers. Water Resources Publications. Littleton, CO. ISBN: 0-477-02608-7. 329 p.

Hubbell D. W. 1987. Bed load sampling and analysis. In: Thorne, C. R.; Bathurst, J. C.; Hey, R. D. (eds). Sediment transport in gravel-bed Rivers. Wiley: Chichester: 89-106.

Ikeda, H.; Iseya, F. 1988. Experimental study of heterogeneous sediment transport. Pap. 12. Environ. Res. Cent. Univ. of Tsukuba. Tsukuba, Japan.

Jackson, W. L.; Beschta, R. L. 1982. A model of two-phase bedload transport in an Oregon coast range stream. Earth Surface Processes and Landforms. 9: 517-527.

Jacobson, R. B.; Coleman, D. 1986. Stratigraphy and recent evolution of Maryland piedmont flood-plains. American Journal of Science. 286: 617-637.

Komar, P. D. 1987. Selective grain entrainment by a current from a bed of mixed sizes: a reanalysis. Journal of Sedimentary Research. 57: 203-211.

Kuhnle, R. A. 1992. Fractional transport rates of bedload on Goodwin Creek. In: Billi, P.; Hey, R. D.; Thorne, C. R.; Tacconi, P. (eds.). Dynamics of gravel-bed rivers. John Wiley & Sons.

Lane, E. W. 1955. The importance of fluvial morphology in hydraulic engineering. J. Hydraul. Div. Am. Soc. Civ. Eng. 81(745): 1-17.

Lisle, T. E. 1995. Particle size variations between bed load and bed material in natural gravel bed channels. Water Resources Research. 31(4): 1107-1118.

McLean, D. G.; Tassone B. 1987. Discussion of Hubbell, DW. Bed load sampling and analysis'. In: Thorne C. R.; Bathurst, J. C.; Hey, R. D. (eds). Sediment transport in gravel-bed rivers. Wiley: Chichester: 109-113.

Meyer-Peter, E.; Müller, R. 1948. Formulas for bed-load transport, Proceedings, 2nd Congress International Association for Hydraulic Research. Stockholm, Sweden: 39-64.

Middleton, G. V.; Southard, J. B. 1984. Mechanics of sediment movement. SEPM Short Course #3. 401 p.

Middleton, G. V.; Wilcock, P. R. 1994. Mechanics in the earth and environmental sciences. Cambridge University Press. 458 p.

Milhous, R. T. 1973. Sediment transport in a gravel-bottomed stream. Ph.D. dissert. Oregon State University. Corvallis, Oregon.

Miller, M. C.; McCave, I. N.; Komar, P. D. 1977. Threshold of sediment motion under unidirectional currents. Sedimentology. 24(4): 507-527.

Mueller, E. R.; Pitlick, J. 2005. Morphologically based model of bed load transport capacity in a headwater stream. J. Geophys. Res. 110: F02016. doi:10.1029/2003JF000117.

Neill, C. R. 1968. A reexamination of the beginning of movement for coarse granular bed materials. Report INT 68. Hydraulics Research Station. Wallingford, England.

Neill, C. R.; Yalin, M. S. 1969. Quantitative definition of beginning of bed movement. J. Hydraul. Div. Am. Soc. Civ. Engrs. 95(HYI): 585-587.

Paola, C.; Parker, G.; Mohrig, D. C.; Whipple, K. X. 1999. The influence of transport fluctuations on spatially averaged topography on a sandy, braided fluvial plain. Numerical Experiments in Stratigraphy. SEPM Special Publication. 62: 211-218.

Parker, G. 2008. Transport of gravel and sediment mixtures. Chapter 3. In: Garcia, M. (ed.), Sedimentation engineering: Processes, measurements, modeling, and practice. Am. Soc. Civil Engineers. Manual 110.

Parker, G.; Hassan, M.; Wilcock, P. 2006. Adjustment of the bed surface size distribution of gravel-bed rivers in response to cycled hydrograph. In: Ergenzinger, P. (ed.). Gravel-bed rivers VI.

Parker, G.; Klingeman, P. C. 1982. On why gravel bed streams are paved. Water Resources Research. 18(5): 1409-1423.

Parker, G. 1990a. Surface-based bedload transport relation for gravel rivers. Journal of Hydraulic Research. 28(4): 417-436.

Parker, G. (1990b) The ACRONYM series of PSACAL programs for computing bedload transport in gravel rivers. External Memorandum M-220. St. Anthony Falls Laboratory, University of Minnesota. 124 p.

Parker, G.; Klingeman, P. C.; McLean, D. G. 1982. Bedload and size distribution in paved gravel-bed streams. Journal of Hydraulic Engineering. 108(4): 544-571.

Parker, G.; Sutherland, A. J. 1990. Fluvial armor. J. Hydr. Res. 28(5).

Pitlick, J.; Cui, Y.; Wilcock, P. 2009. Manual for computing bed load transport in gravel-bed streams. Gen. Tech. Rep. RMRS-GTR-223. Fort Collins, CO: U.S. Department of Agriculture, Forest Service, Rocky Mountain Research Station. 40 p.

Reid, I.; Frostick, L. E.; Layman, J. T. 1980. The continuous measurement of bedload discharge. J. Hydraul. Res. 18: 243-249.

Reid, L. M.; Dunne, T. 2003. Sediment budgets as an organizing framework in fluvial geomorphology. Ch. 16. In: Kondolf, G. M.; Piegay, H. (eds.). Tools in fluvial geomorphology. John Wiley & Sons.

Reid, L. M.; Dunne, T. 1996. Rapid evaluation of sediment budgets. Catena Verlag. Reiskirchen, Germany. 164 p.

Rosgen, D. 2007. Watershed assessment of river stability and sediment supply (WARSSS). Wildland Hydrology.

Schmidt, J. C.; Wilcock, P. R. 2008. Metrics for assessing the downstream effects of dams. Water Resour. Res. 44: W04404. doi:10.1029/2006WR005092.

Trimble, S. W. 1998. The use of historical data in fluvial geomorphology. Catena. 31: 283-304.

Wilcock, P. R. 1988. Methods for estimating the critical shear stress of individual fractions in mixed-size sediment. Water Resources Research. 24(7): 1127-1135.

Wilcock, P. R. 1996. Estimating local bed shear stress from velocity observations. Water Resources Research. 32(11): 3361-3366.

Wilcock, P.R. 1993. The critical shear stress of natural sediments. The Journal of Hydraulic Engineering. 119(4): 491-505.

Wilcock, P. R. 1997. Entrainment, displacement and transport of tracer gravels. Earth Surface Processes and Landforms. 22: 1125-1138.

Wilcock, P. R. 1998. Two-fraction model of initial sediment motion in gravel-bed rivers. Science. 280: 410-412.

Wilcock, P. R. 2001. Toward a practical method for estimating sediment-transport rates in gravel-bed rivers. Earth Surface Processes and Landforms. 26: 1395-1408.

Wilcock, P. R.; Crowe, J. C. 2003. Surface-based transport model for mixed-size sediment. Journal of Hydraulic Engineering. 129(2): 120-128.

Wilcock, P. R.; DeTemple, B. T. 2005. Persistence of armor layers in gravel-bed streams. Geophysical Research Letters. 32: L08402. doi:10.1029/2004GL021772.

Wilcock, P. R.; Kenworthy, S. T. 2002, A two fraction model for the transport of sand-gravel mixtures. Water Resources Research. 38(10): 1194-2003.

Wilcock, P. R.; Kenworthy S. T.; Crowe, J. C. 2001. Experimental study of the transport of mixed sand and gravel. Water Resources Research. 37(12): 3349-3358.

Wilcock, P. R.; Kondolf, G. M.; Matthews, W. V. G.; Barta, A. F. 1996. Specification of sediment maintenance flows for a large gravel-bed river. Water Resources Research. 32(9):2911-2921.

Wilcock, P. R.; McArdell, B. W. 1993. Surface-based fractional transport rates: mobilization thresholds and partial transport of a sand-gravel sediment. Water Resources Research. 29(4): 1297-1312.

Wilcock, P. R.; McArdell, B. W. 1997. Partial transport of a sand/gravel sediment. Water Resources Research. 33(1): 235-245.

Williams, G. P.; Wolman M. G. 1984. Downstream effects of dams on alluvial rivers. Professional Paper 1286. U.S. Geological Survey.

Wong, M.; Parker, G. 2006. Reanalysis and correction of bed-load relation of Meyer-Peter and Müller using their own database. J. Hydr. Engrg. 132(11): 1159-1168.

# Appendix—List of Symbols

## Dimensioned Variables

| Symbol | Definition | Units | Key location (Eq. or chapter) |
|---|---|---|---|
| $a$ | coefficient in sediment rating curve | *variable* | 2.4, 6.1-6.4 |
| $b$ | exponent in sediment rating curve | none | 2.4, 6.1-6.4 |
| $B$ | channel top width | L | 7.2 |
| $c$ | coefficient in generic transport formula | none | 2.20 |
| $d$ | exponent in generic transport formula | none | 2.20 |
| $D, D_i$ | grain size, subscript $i$ for size fraction $i$ | L | |
| $D_s, D_{gr}$ | mean grain size of the sand and gravel fractions in a two-fraction approach | L | Chapter 2 |
| $D_{xx}$ | grain size for which $xx$ percent of the bed is finer | L | |
| $F_i$ | proportion of size fraction $i$ on the bed surface | none | |
| $f_i$ | proportion of size fraction $I$ in the bed subsurface | none | |
| $g$ | acceleration of gravity | $LT^{-2}$ | |
| $h$ | flow depth | L | |
| $k$ | coefficient in velocity rating curve | *variable* | 3.2 |
| $m$ | exponent in velocity rating curve | none | 3.2 |
| $n$ | Manning's roughness coefficient | $TL^{-1/3}$ | 2.9, 7.3 |
| $n_D$ | Manning's roughness due to bed material only (Manning-Strickler) | $TL^{-1/3}$ | 2.11, 4.4 |
| $p_i$ | proportion of size fraction $i$ in transport | none | |
| $Q$ | water discharge | $L^3T^{-1}$ | |
| $Q_c$ | critical water discharge for incipient sediment transport | $L^3T^{-1}$ | Chapter 7 |
| $Q_r$ | water discharge at reference transport conditions | $L^3T^{-1}$ | |
| $Q_s$ | sediment transport rate | *various* | 2.6, 6.1-6.4 |
| $q$ | water discharge per unit width | $L^2T^{-1}$ | |
| $q_s, q_{si}$ | sediment transport rate per unit width, where subscript $i$ refers to an individual size fraction $i$ | $L^2T^{-1}$ | 2.16 |
| $R$ | hydraulic radius | L | 2.5 |
| $S$ | channel slope | none | |
| $S_f$ | friction slope (slope of the energy grade) line) | none | 2.7 |
| $t$ | time | T | 2.6 |
| $U$ | mean velocity | $LT^{-1}$ | |
| $w$ | sediment fall velocity | $LT^{-1}$ | Chapter 2—Dimensional analysis |
| $x$ | distance in streamwise direction | L | 2.6 |
| $z_b$ | streambed elevation | L | 2.8 |
| $\phi$ | grain size scale, for $D$ in mm $\phi = -\log_2(D)$ and $D = 2^{-\phi}$ | | Chapter 2—Grain size |
| $\mu$ | water viscosity | $ML^{-1}T^{-1}$ | 2.16 |

# Dimensioned Variables (cont.)

| Symbol | Definition | Units | Key location (Eq. or chapter) |
|---|---|---|---|
| $\rho$ | water density | $ML^{-3}$ | |
| $\rho_s$ | sediment density | $ML^{-3}$ | |
| $\sigma_g$ | geometric standard deviation of grain-size distribution | | 2.3 |
| $\sigma_\psi$ | arithmetic standard deviation of grain-size distribution on the $\psi$ scale | | 2.3 |
| $\tau$ | shear stress | $ML^{-1}T^{-2}$ | |
| $\tau'$ | grain stress | $ML^{-1}T^{-2}$ | 2.15, 4.4 |
| $\tau_0$ | boundary shear stress | $ML^{-1}T^{-2}$ | 2.5 |
| $\tau_c, \tau_{ci}$ | critical shear stress for incipient grain motion where subscript $i$ refers to an individual size fraction $i$ | $ML^{-1}T^{-2}$ | 2.21, 2.30 |
| $\tau_l$ | local shear stress | $ML^{-1}T^{-2}$ | Chapter 3—The Flow Problem |
| $\tau_r, \tau_{ri}$ | reference shear stress, surrogate for $\tau_c$ | $ML^{-1}T^{-2}$ | 2.25 |
| $\omega$ | straining function in Parker (1990) transport function | | 4.2 |
| $\psi$ | grain size scale, for $D$ in mm $\psi = -\log_2(D)$ and $D = 2^{-\psi}$ | | Chapter 2—Grain size |

# Dimensionless Variables

| Dimensionless group | Definition | Location |
|---|---|---|
| $q*$ | $\dfrac{q_s}{\sqrt{(s-1)gD^3}}$ | 2.18 |
| $s$ | $\rho_s/\rho$ | 2.18 |
| $S*$ | $S* = \dfrac{\sqrt{(s-1)gD^3}}{\mu/\rho}$ | 2.18 |
| $\tau*$ | $\dfrac{\tau}{(s-1)\rho g D}$ | 2.18 |
| $W*$ | $\dfrac{(s-1)gq_s}{(\tau/\rho)^{3/2}}$ | 2.23 |

Subscripts on $\tau*$, $q*$, and $W*$:

    $c$: critical conditions for incipient sediment motion

    $i$: applies to individual grain size fraction

    $r$: conditions at reference sediment transport rate ($W* = 0.002$)

    $g, s$: applies to gravel or sand fraction in two-fraction approach

    50: median grain size